THE AGE OF ECOLOGY

THE AGE OF ECOLOGY

The Environment on CBC Radio's Ideas

David Cayley

James Lorimer & Company, Publishers
Toronto, 1991

QH 540.5 . C38 1991

Copyright ©1991 James Lorimer and Company, Publishers

All rights reserved. No part of this book may be reproduced or transmitted in any form or by any means, electronic or mechanical, including photocopying, or by any information storage and retrieval system, without permission in writing from the publisher.

Canadian Cataloguing in Publication Data

Cayley, David
    The age of ecology

ISBN 1-55028-351-0 (bound) ISBN 1-55028-349-9 (pbk.)

1. Ecology - Philosophy. 2. Human ecology.
3. Environmental policy. I. Title.

QH540.5C38 1991   304.2'8   C91-094853-4

Cover photo: David Laurence

James Lorimer & Company, Publishers
Egerton Ryerson Memorial Building
35 Britain Street
Toronto, Ontario
M5A 1R7

Printed and bound in Canada

Contents

Acknowledgements		v
Introduction		vii
I	New Ideas in Ecology	1
1	Deep Ecology	3
II	Citizens at the Summit	19
2	Environment, Development and Debt	21
3	"You Have to Keep Swimming"	41
III	From Commons to Catastrophe	53
4	The Last Assault	55
5	The People of the Forest	71
6	Preserving the Forest	93
IV	The Age of Ecology	113
7	A Managed Planet	115
8	One-Eyed Science: A Conversation with Vandana Shiva	134
9	Ecology as Design: The Work of John Todd	144
10	Gaia: A Way of Knowing	163
11	David Ehrenfeld: Stewardship and the Sabbath	183
V	Redefining Development	195
12	A Mental Ruins	197
13	Development and Democracy	216
14	The Steady State	237
15	The Ambivalence of Ecology	258
	Suggestions for Further Reading	269

Acknowledgements

The following book comprises transcripts of radio programmes broadcast on CBC Radio's *Ideas* series between 1986 and 1990. I am indebted to executive producer Bernie Lucht, producers Sara Wolch and Jill Eisen who worked on some of these programmes with me, and to my other colleagues at *Ideas* for their support and encouragement.

Radio documentary is very much a found, rather than a created, art; and broadcasters, in consequence, depend heavily on the kindness, and indulgence, of strangers. Many of the people quoted in these pages had never heard of me when they received a phone call out of the blue, requesting an interview and offering as evidence of my good intentions nothing more than a hasty, and quite possibly garbled, account of what I had in mind. Without their courteous and often generous responses, there would have been no programmes. Since they are too numerous to mention individually, and all are named in what follows, I thank them collectively here.

Finally, I would like to thank Carolyn MacGregor and Jim Lorimer, who convinced me that these transcripts were not just accessories to the original broadcasts, but deserved an independent literary existence.

Introduction

The summer of 1988 was hot and dry. Day succeeded cloudless day, and still there was no rain. The ground dried and cracked, and the corn in the fields looked scrawny and stunted. It might have just been an aberration, a drought like other droughts; but, during that summer, a scientific congress of climatologists assembled in Toronto suggested that this drought had something uncanny and unprecedented about it. The scientists' opinions were divided, and the proofs they provided were all contested, including even the possibility of proof; but there were many who were prepared to state that the weather outside was a sign of a major change in the biosphere, a warming of the entire atmosphere brought about by human activities.

The idea was not a new one. It had occurred to the scientist Jean Baptiste Fourier as a purely logical deduction two hundred years ago: transform the carbon stored in the earth into atmospheric carbon dioxide by burning carbon fuels, and sooner or later this blanket of heat-trapping gas will warm up the earth like a greenhouse. At the end of the nineteenth century, Svante Arrhenius, a Swedish scientist, actually tried to calculate the precise effects of "evaporating our coal mines into the air." In the 1970s, articles had begun to appear in the popular press that mapped the possible inundation of the low-lying coastal areas, should the polar ice-caps begin to melt. But this remained speculative. In Toronto, in 1988, it suddenly seemed frighteningly real.

Ecological crises have been recurrent throughout human history. Human beings have been sculpting, and sometimes ruining, their landscapes since the late neolithic era; but the greenhouse effect represents something unprecedented: a change so colossal that it registers in the great scale of geological time. Other civilizations have collapsed after stripping their forests and despoiling their soils, but these have been essentially local disturbances from which nature has always recovered. Today the pressure on nature is intense and unremitting, and there is neither time nor space left for natural regeneration.

We live at an apocalyptic moment, a moment when the fate of our civilization is revealed, and fundamental choices are posed. The prediction of eco-catastrophe is now so commonplace that it has begun to lose its power to shock. In places like Haiti, Nepal and the Sahel, where the land is already stripped and bare, it is no longer a prediction: the catastrophe has already taken place. But possible responses move in two very different directions: one towards humility and one towards a new form of the hubris which has brought us to the edge of catastrophe in the first place. By *humility* I mean the response that seeks a more modest and more fitting way of life and not just technical adjustments in the means of production. It is a response that discerns in the environmental crisis a deeper question than merely how to survive, a question about the meaning of the existence which surrounds us and a question about our proper relationship to this existence. By *hubris*, I mean that response which proposes a more prudent, and at the same time, more intensive management of "resources" and intends to carry out this management on a planetary scale. This response has produced a new class of activists and thinkers whom German thinker Wolfgang Sachs dubs "ecocrats" — people who think that the crisis can be managed by throwing a net of ecological expertise over the entire planet.

During the ecological panic of the late 1980s, this new class of "experts" came very much to the fore. Typical was the September 1989 issue of *Scientific American*, which announced on its cover the theme "Managing Planet Earth." "It is as a global species that we are transforming the planet," wrote William Clark in the lead article. "It is only as a global species, pooling our knowledge, coordinating our actions and sharing what the planet has to offer, that we have any prospect for managing the planet's transformation along pathways of sustainability." This statement encompasses the whole earth in a single vision. Boundaries are erased in favour of a presumably homogeneous global species. Cultural and religious horizons dissolve. The ecological crisis, in Clark's eyes, demands neither renunciation nor virtuous action; it demands expert management.

The emblem of this new mood of globalism, the sign of its piety as well as its panic, is the image of the earth photographed from space. It is an image that conveys all the ambiguity of the new age. The tiny blue-green marble in its mantle of cloud suggests something vulnerable and finite, but it also suggests something small and manageable, something that can be comprehended at a glance, something that invites control even as it fills us with pseudo-mystical emotion. It appears to be a natural object, and yet it is utterly alien to our sensed experience of earthly existence. Without the vast militaristic apparatus of space exploration, it could not exist at all. The view from the moon is remote, distant and detached. The earth it visualizes is a technological artefact, an abstraction masquerading as an object of common sense.

Ecology is like the Roman god Janus, the presiding deity of gates and doorways, with his two heads facing in opposite directions. One head gazes with the imperious eye of technocratic science, eager to manage the planet. The other looks back nostalgically on an order of Nature to which we belong but which cannot belong to us. This contradiction, as historian Donald Worster points out in these pages, was part of ecology even before biologist Ernst Haeckel named the new science in 1873. Worster calls these two faces of ecology the arcadian and the imperialist. The arcadian tendency, with its awe of nature and its protest against the utilitarian ethos of industrialization, has its origins in the counterculture of Romanticism. The imperialist tendency is conventionally scientific, favouring prediction and control and reducing living nature to a set of bloodless abstractions. These tendencies were mixed in the environmental movements that arose in the Western countries at the end of the 1960s. Philosophical schools, political movements and workaday science all shared the name of ecology. However, as the crisis deepened, and environmentalism moved more into the mainstream, the contradictions began to be exposed. The romantic and visionary side of ecology was gradually eclipsed, and the managerial side became more prominent. By the late 1980s everyone was an environmentalist, green products sold well at the supermarket, and the environment had become one more calculation within the field of administrative technique.

That this would eventually happen had been predicted by prescient writers from the very beginnings of contemporary environmentalism. Ivan Illich, speaking in Toronto in 1970, described the degradation of nature as "the result of a corruption in man's self-image." This corruption, he said, consisted in the growing intermediation of institutional services both between people and between people and nature. The alternative he proposed was self-limitation. Otherwise, he said, limitation would be imposed and eventually transform the environment into a new "service sector" within the economy. The prospects for freedom through renunciation would be foreclosed, and even the minutest gestures of daily life would have to be weighed, monitored and measured for their environmental impact. In 1980, in his book *Ecology As Politics*, Andre Gorz made a similar assessment. Ecology, he said, can be a tool either of social transformation or of intensified management: pollution and environmental degradation will either generate new forms of popular sovereignty, or they will become the excuse for a more stringent capitalist hegemony.

Today, when the Brundtland Commission proposes that only "a new era of economic growth" can reduce pressure on the environment, and the Bush administration addresses the problem of acid rain with marketable pollution quotas, it's clear the latter path has been chosen. The radicals still exist, but the range of choices narrows with each day they are unheeded, and ecology becomes more and more the creature of the existing institutions.

The bulk of popular writing on the environment at the beginning of the 1990s seems to fall into two main genres. These can be caricatured as the Chicken Little approach, which warns that the sky is falling, and the Pollyanna approach, which suggests saving the planet by putting new ringers in your leaky faucets and shopping more wisely. The distance between the severity of the perceived problems and the triviality of the proposed solutions demonstrates how completely social, economic and philosophical questions have disappeared from discussions of the environment. This distance is the difficulty that this book tries to address. The more urgent the environmental problem becomes, the greater is the tendency to shoot first and ask questions later. I want to suggest here that no matter how urgent the problem is, we still need to approach ecology in a quizzical, inquiring manner that seeks first to grasp the difficulty, before devising a solution. The alternative is to try to make the problem fit the solutions we already have, and let the deeper, more philosophical questions remain unanswered.

Let me give an example of what I mean. It has been intuitively obvious to many people for at least twenty years that any adequate approach to the environmental crisis would involve serious shrinkage in the scale of economic activity, as well as in the proportion of people's lives dominated by formal economic relations. It is also clear that even the mildest recession in economic growth involves painful suffering for the weakest members of society. So between what we would like — a virtuous relationship with nature — and what we are committed to — social justice through economic growth — there is a gulf, and honest discussion can only begin by acknowledging this separation.

Another more specific example is the current controversy over the second phase of hydroelectric development in James Bay. The native peoples speak against the flooding of their homeland. They justify their opposition on the grounds that they have lived there for a long time and this is not how they want their land to be used. The government of Quebec argues in the utilitarian logic of jobs, money, and the greatest good for the greatest number. There is no common coin into which these two views can be converted: they are completely incommensurable. In modern societies, as Alastair McIntyre has argued in his book *After Virtue*, moral views are always disguised preferences. They can argue but they can never agree. In this case, a problem in ecology suddenly turns into a conundrum in moral philosophy.

There is currently a good deal of talk about *sustainable development*, a term popularized by the United Nations' Brundtland Commission. It means, roughly, using the environment in such a way that its future usefulness will not be impaired. The Commission suggests that preserving the environment is just sound business practice. This seems to offer an answer, but in fact it just forecloses the question. With sustainable development, nature is fully incorporated within the categories of economics. Where once nature sustained various human economies, now the economy sustains nature and justifies its continued existence. This is what lies behind the Commission's paradoxical claim that economic growth is now the only way to preserve the environment. The critique of economics which ecology at first seemed to imply has become instead the final triumph of economics.

If politics is the art of the possible, and what is possible for a given civilization is largely determined by the ideas and institutions it has established over centuries, then it can be persuasively argued that sustainable development is the only answer the modern West can give to the question of ecology. It was a political maxim of Sir Thomas More's that where the best is impossible, one should try to prevent the worst. On these terms, sustainable development appears, at the least, to be a way of preventing the worst of the current abuse of nature. Why dwell in fruitless misgivings, and what George Grant called "intimations of deprival"? Why occupy ourselves with older ideas of nature that patently don't fit the contemporary economic calculus to which the entire world must now inevitably submit? Why not just adjust our consumption to our best calculation of what nature will yield in perpetuity, and forget about the painful ironies and unanswered questions to which we are otherwise exposed?

My response is that, if a society conceives of nature as a productive system, and its citizens as consumers of resources, then it finally forsakes the question to which all previous cultures have given their imperfect answers: What does the existence of the world mean, and what, in the light of this meaning, is it good for human beings to do? The sentences quoted earlier from William Clark make plain how the currently dominant discourse of ecology makes it impossible even to pose this question. In Clark's terms we are no longer cultural beings at all. We are a "global species," and, as such, we give ourselves the task of "managing the planet." There is no sense that the world is not ours to manage, no reason to wait on things until they declare their meaning to us. The world has no meaning apart from our inhabitation of it, and there is nothing to guide us except prudential and utilitarian calculations about "what the planet has to offer." Distinct peoples, places and traditions all dissolve into the homogeneous existence of the shrunken and domesticated globe.

*** *** ***

The programmes assembled here are drawn from five different series made between 1986 and 1990 for CBC Radio's *Ideas* series, where I work as a writer/broadcaster. They tell a double story: the story of my growing awareness of the gravity of the environmental crisis, and the parallel story of my increasing misgivings about the way in which we conventionally think and speak about it. The book begins with a thumbnail sketch of "deep ecology," a branch of environmentalism which is attempting to prod the movement into a more philosophical mood. The two sections that follow, "Citizens at the Summit" and "From Commons to Catastrophe: The Destruction of the Forests," describe the causes and consequences of environmental degradation in the countries of Asia, Africa and Latin America. They emphasize the indissoluble link between culture and environment and point to the fact that environmental solutions must always first be cultural and political solutions. "The Age of Ecology" returns to philosophical themes and tries to undermine current certainties by exploring some of the contradictory tendencies within environmental thought. The concluding section, "Redefining Development," looks at how development has destroyed subsistence and asks whether "sustainable development" can ever be more than a contradiction in terms.

Each series, as I have tried to make clear in the introductions, was made in response to a particular occasion, and there are inevitable repetitions and discontinuities that are impossible to eliminate. What they have in common is an insistent question which facile responses to the environmental crisis often obscure: a question not about what is sustainable but what is good.

The original transcriptions of these programmes published by the CBC are verbatim records of what was broadcast. I have sometimes shortened programmes and made other alterations in the interests of a more compact book. I have also replaced the programme openings which I originally wrote for *Ideas* host Lister Sinclair with new introductions which attempt to set each section into its new context. I offer these programmes not as finished arguments but as fruitful starting points.

I
New Ideas in Ecology

1

Deep Ecology

"New Ideas in Ecology and Economics" was broadcast in May and June, 1986. For years, I had been following a variety of hopeful ideas and initiatives in these fields, and I finally decided to track some of them down. My journey took me from the New Alchemy Institute on Cape Cod, where I saw buildings that produced food and regulated their own climates, to the Mattole Valley of California, where the residents were trying to restore a watershed ruined by careless logging. I canvassed the emerging practice of "alternative economics" and looked at community economic development corporations from Sydney to Nanaimo. The general idea was to integrate the study of ecology and economics, and describe the ways in which this new synthesis was finding embodiment.

The programme that is featured here examines a critique of mainstream environmentalism that is sometimes called "deep ecology." Deep ecology is a term coined by Norwegian philosopher Arne Naess. He first used it in the early 1970s when he published an influential essay that distinguished between the two types of ecology, deep and shallow. The distinction is so invidious that Naess's followers have subsequently tended to speak of "reform" rather than "shallow" ecology, but it did describe a real rift in environmental thought between those interested in administrative solutions to environmental problems and those interested in uncovering the political, economic, and cultural roots of these problems. Finding technocratic fixes for environmental problems, reasoned the deep ecologists, was like cutting off the heads of the mythical Hydra: where one was severed, two would grow back.

In Canada one of the voices of deep ecology is a journal published in Victoria called *The Trumpeter*. Its editor is Alan Drengson, a teacher of philosophy at the University of Victoria, and he began the programme.

Alan Drengson
When Naess was asked in an interview if he could state the essence of deep ecology as he understood it, he said, think of it in terms of an inquiry that you're carrying on. He says, of course, you could ask what is the source of a particular environmental problem, say, of a certain pollutant in a river. You'd follow that river back up and you'd find a certain factory that makes antifreeze and it releases an effluent, et cetera. And you stop there. Now that's a reform, or what Naess called a shallow ecological approach. You simply stop there.

But Naess said what we do as deep ecologists is go on to ask what form of education, what form of religion, what form of politics, what form of economics, what forms of technology, and so on, are most conducive to the integrity and the well-being of all beings of planet Earth. Not just human beings, but all beings. And this kind of inquiry that Naess is talking about is an ever-deepening questioning of our ends, our values. In short, these are the deep questions that philosophy has always asked, but what Naess was saying is that now we have to ask them in the ecological context with the ecological consciousness that is emerging and is being developed as a result of the efforts of all kinds of different people, not just philosophers, but scientists and artists and so on.

David Cayley
Alan Drengson insists that deep ecology is not a set of fixed principles. It's an approach, an attitude. Essentially, although the term is overused, it's a holistic philosophy. It says that the whole is greater than the sum of its parts and that we realize value in our lives through our relatedness to this larger whole.

Alan Drengson
With respect to the philosophy of nature, what the deep ecologists seem to say is that we find our sense of well-being and worth not by withdrawing into subjectivity, nor by denying values that exist in the world, but by involving ourselves in an active way with the natural world, with our human communities. It's through our relationships that we realize value. So what they're saying is that if you withdraw into yourself alone, in total subjectivity, all values seem to go. That seems to be a kind of nihilism. And the way out of that is to become involved with others, other people, but the human species as a whole could still lack a sense of grounding of values unless it makes peace with the natural world. Nature is not engaged in warfare against humans, it's human beings that are engaged in a kind of warfare against nature. We have to come to recognize that there's more to self than narrow subjective human self, and that the more we recognize this, the more deeply we experience this, the more meaningful our lives become. In some respects, deep ecology is a response not only to environmental degradation, but to a very deep problem in modern human consciousness.

David Cayley
To get outside of our subjectivity, Drengson says, we should have direct contact with non-human nature. To him, this means experiencing wilderness.

Alan Drengson
If you go back to the very roots of the word *wilderness*, what you find is that you can rephrase the word to be something like "will of the land." Ancient cultures, for instance, the Celtic cultures, the Roman culture, the earlier Greek culture, particularly those cultures that were centred around oral traditions before literacy became more widespread, thought of the land as having a will of its own. In the Roman context, they talked of it in terms of the spirit of the place. Before you could build in a place, you had to divine its spirit, so to speak, and determine its will. In a sense, the will of the land refers to the fact that nature exists in its own right for itself. It existed before humans came along and will more than likely exist after humans leave the planet, unless they totally destroy it. Also,

other beings have their own ways, their own will, and in wilderness, especially in big wilderness, nature is allowed to continue its evolutionary destiny without human manipulation. So in that respect, you could say that by going back to wilderness, we rediscover our own wild nature that's in each and every one of us.

I mean, think about how much our lives are dominated by clock time, which is a totally mechanical, arbitrary form of time-keeping. If you go into the wilderness and you don't take clocks with you, and just live by natural rhythms, you find that after a few days, something rather interesting happens to your sense of time. Your time consciousness is quite different in the wilderness than it is in the city. The city is structured around the clock. A modern industrial city is in part a product of a clock, and without the clock, it couldn't exist. If all our clocks quit working tomorrow, it'd be chaos. I don't know if it'd collapse, but it'd certainly be strange.

So, people see wilderness in a lot of different ways. It's something in itself that has its own will, its own way, its own destiny, but also it's something that is in each and every one of us. No matter how much culture has been piled on top of it, there is a natural human being underneath there somewhere.

David Cayley
Deep ecology proposes a return to nature, but this doesn't mean turning away from modern science and going back to a more primitive world view. Alan Drengson sees it as an attempt to reach a higher type of science through what he calls "a new apprenticeship to nature."

Alan Drengson
Deep ecology is an inquiry that attempts to bring us to a much deeper understanding of nature with all its mysteries and strangeness, but also nature's reciprocity with ourselves. It is a return to an objective foundation of value for our lives. That objective foundation is the entire matrix of the natural world.

Maybe I can put this in another way with an example of natural farming. The natural farmer attempts to farm with a minimum disturbance of the natural world and attempts to maintain as much habitat as possible. The disruption of biological communities is minimized in that kind of agriculture. In connection with deep ecology this represents an apprenticeship to nature through which the farmer has learned to farm the way nature farms. To me, this represents the deepest kind of science, because if science is an inquiry whose aim ultimately is to understand the way the world is in itself, then that kind of observation of the natural world that brings us the deepest understanding is the most scientific. People are beginning to realize that approaching the world with a certain kind of compassion brings a deeper understanding than does approaching it as an alien being to be feared or conquered.

David Cayley
One of the terms which deep ecology has posed as a problem is *environment*. Unlike nature, environment is a purely relational term, a something which surrounds an implied something else. Lacking self-identity, or the possibility of reciprocity, environment is always implicitly "our environment." The term is investigated and challenged in a book called *The Natural Alien* by Neil Evernden, a zoologist in the Faculty of Environmental Studies at York University.

Neil Evernden
When you speak of environment, you automatically presuppose that all the value, all the life part has been scraped off into the important viewing subject. It's like a Renaissance painting where the viewer looks out across an empty world, surveying it in a detached way, and then you've got environment. Once there's no involvement, once there's no interaction with it, it's just "that stuff." And to describe someone as an environmentalist, then, is to describe someone who is interested in "that stuff." And that in turn is misleading, because what led that individual to this concern was not "that stuff," but his or her relationship to it, his or her involvement with it as a field of self. One's feeling of being a part, of being at home, in being involved, in being in the world — in being. So ironically, the very concept of environment reveals an attitude that prevents one ever being part of ever accomplishing what the original motivation was, namely a personal, deep commitment to place.

David Cayley
To Neil Evernden, the word *environment* describes what's "out there," apart from us. It's a word with a whole philosophy hidden inside it — Cartesian science. Cartesian science began in the seventeenth century, when the French philosopher René Descartes split the world into subjects and objects.

Neil Evernden
Cartesian science gives you a means of accomplishing something, but it's a mental trick requiring that you first forget something. It's a trick that divides experience into two categories, that which can be handled and that which can't. Now, that which can be handled, the quantitative, it does very well with. The danger is that after a while it begins to seem like that's all there is, or at least all that is worth thinking about. What happens if something of concern to you arises that does not fit exclusively into that category? You're left without any forum to talk or any tools to speak with.

And I think that's where the problem comes in, in the reliance we have on modern science, the unspoken assumption that only if it can be discussed in scientific terms is it worth saying. That's another of the encouragements to translating concerns into materialist terms, of course, but more often it's a barrier to ever speaking about the experience that moves you in the first place, because it almost seems by definition to be unreal. It is by definition unreal if you want to be fundamentally accurate about it in scientific terms. One's experience is considered subjective and therefore not real in the sense that we usually use the term. So the personal experience that moves the environmentalist in the first place is the unreal part, which you don't talk about. Instead, you talk about the material things that seem to be the expression of this concern, like polluted water. Not the experience of the person, not the significance of the water, but something you can measure in water.

David Cayley
Neil Evernden has described a kind of Catch-22. We may feel deep concern, even love, for nature, but when we translate those feelings into scientific language, we only betray them. John Livingston has described a similar problem. Livingston's

a life-long naturalist and a colleague of Neil Evernden's at York University's Faculty of Environmental Studies. In 1981, he published an essay called "The Fallacy of Wildlife Conservation." It read like a cry of despair. A career in wildlife conservation had convinced Livingston that conservationists could only get a public hearing if they based their arguments on the usefulness of nature to human beings. This made conservationism palatable to people, but in Livingston's view, it also undermined the movement itself. In trying to make arguments against the human domination of nature, one only ended up reinforcing it.

John Livingston
I felt that I was looking at something that had developed since the early part of the century, where wildlife was seen entirely in utilitarian terms. In order to justify the preservation of species or of habitats of groups of species, you had to justify them in some sort of utilitarian way, and thereupon use the term *resource* to save something that obviously has no utility except its very existence.

David Cayley
Can you describe some of these utilitarian rationales that are put forward?

John Livingston
Well, we must not lose — *we* must not lose, note that — we must not lose the Amazonian rain forest, for example, because the Amazonian rain forest supplies x percentage of the oxygen to our atmosphere. Point number one. Point number two is that we must not destroy those forests because who knows what goodies may be hidden there that may be of use to us some day? There may be a tree or a shrub that will produce some highly desired cosmetic, or there may be a tree or shrub or an animal that provides some much desired poison. In Florida, I once saw a sign outside a shell emporium saying "Rarities unknown to exist." And this is much the same thing. The Amazonian rain forest is filled with rarities and wonderful goodies and things for us that we don't know about yet, therefore it must be conserved. But this is as far as we've been able to get.

David Cayley
So why can't we get any further? Livingston believes it's because our whole culture denies nature an independent existence. We believe nature's there to serve us.

John Livingston
To make any argument, any rational, linear, logical argument on behalf of the whooping crane is impossible in our language, and it's impossible also within the overarching system of beliefs that we've inherited. Because the thing isn't of any value. It's probably worth $1.98 if you were to weigh it on some scale. But is that the point? Must we marshall logical argument in order to address that which is not logical? The existence of the whooping crane is not logical.

David Cayley
Everything John Livingston and Neil Evernden have been saying points to a paradox. Our separation from nature frustrates and subverts our attempts to preserve it. Their writings take this further and suggest that we'll never solve our environmental problems until we find a language, a philosophy and a way of life which overcomes this separation, and this means trying to unearth the roots of our alienation. Unfortunately for John Livingston, alienation virtually defines human culture.

John Livingston
The overwhelming human speciality is transmissible technique. Transmissible down the generations. That's what culture is, transmissible and transmitted techniques of doing things. The most important technique in our history, in my view, has been how to exercise social control. Everything else thereafter became utterly extraneous. Everything else became either meat for the fire or water, or whatever, and was serving a means of thought and a means of living that had utterly transcended our biological nature. We lived in a fabricated world long before we began to fabricate things of any particular significance, because it was an intellectually fabricated world.

David Cayley
Why does the practice of social control denature nature?

John Livingston
Because it invites the world to be contained within the invisible sphere that represents that society. It's like what Northrop Frye says about myth. I believe he said that one of the functions of myth was to set a stockade around a society and prevent it from having any further truck or trade with nature on the outside; to contain itself. That's what I mean by an invisible bubble around a society. The more complex, the more sophisticated the social rituals and the social constructs become, the more nature is denatured and the more meaningless nature becomes.

David Cayley
Overcoming this isolation, says Livingston, means getting outside of our limiting cultural assumptions, and the only way we can do that is by recognizing that we are also members of the larger community of nature.

John Livingston
What we're looking for is not so much some recipe for the future, but a recovery of that which already exists in us. My hope, all of my optimism, rests in the fact that I bleed when I'm pricked and I bruise when I'm hit, and thank God I'm a biological being. And since I am a biological being, I still have in me that wondrous capacity to live as part of a greater enterprise than I, and a greater enterprise than that of my own species and my own family. I have evolved to be integrated with and part of and able to participate in something that's much grander and more continuing than my species.

All animals, I will assert, are largely if not almost entirely unaware of self and self-interest. Interspecies, that is to say, actions of animals across species lines, between different kinds of animals, all seem to me to demonstrate a mutual and reciprocal and cooperative level of mutual understanding that is quite difficult to find even between two human individuals of the same species, let alone across species. I believe that this exists — I know that it exists from my observation of animals. I've done little else in my life but watch animals.

I know that this recovery of my animalness is possible. It's difficult for me to put it into words because words and language have a tyranny of their own, and so do metaphors and so forth. But I do believe that from looking at natural communities and the way they seem to work, and the way they seem to work in such a wonderful reciprocal way, that there is a chance for us to simply try and recover that which we have amputated. I believe that we are a prosthetic being. Our cultures and traditions are prostheses set in place of that which we have conceptually amputated, and to throw away the crutch that's represented by our cultural beliefs, realities and so on, would be very difficult.

David Cayley
When you speak of cooperation, what do you mean?

John Livingston
I mean effective cooperation. I mean that the net result could only have occurred had there been mutual participation. Whether it's chasing and catching something, or finding something, or feeding, or whatever. Various people, Barry Lopez being the best known, have postulated something he calls the conversation of death. I see it rather as the conversation of life between the wolf and the caribou, or between the wolf and the deer. I have seen this myself. The deer identifies itself as the one that's going to be consumed on that day. That's not just effective cooperation, that's utter mutualism, in my view. This happens in nature daily. Instant by instant.

David Cayley
John Livingston also believes that this kind of participation is available to human beings. He says he experienced it himself one day, while he was diving.

John Livingston
I'm a great swimmer and coral reef snooper, and I've done this and had a marvellous time. We love to do this whenever we have a chance. I once lined up in the cleaning station — you must have heard of the cleaning station. The small young of many species of fish and even the adults of some species of saltwater fishes clean other fishes, make their living by eating parasites off them the way tick birds eat parasites off elephants and so on. And it's like a car wash, and you line up. And I recall lining up behind a nice, big grouper, lined up behind

him and got cleaned. There was a parrot fish lined up behind me. You just go through the lineup, then they clean all the stuff off you, and away you go, exactly like a car wash. And that I find is mutualism in the extreme. They're there to do their stuff, you come and do it, and you've done it together, and everybody benefits. In nature, everybody benefits.

David Cayley
Not everyone can line up at a fish "car wash," but we can all have some kind of direct experience of nature, and it's only through such personal experience that Livingston says we'll ever gain a new attitude towards other forms of life.

John Livingston
Such an attitude can only come from individual experience of the non-human. And I don't care if it's a poor little fish in a goldfish bowl, or a little potted geranium on the window sill, or the woods outside my window here, or my dogs, or whatever it is. The child, and the mature individual too, must have contact with that which is not human and which is not of human manufacture. I believe that we live in a society of chronic sensory deprivation, not in the classical psychological sense, but in the sense that we are deprived of the stimuli that can only derive from sources which are not human and not of human manufacture. Whereas every non-human being is born and grows up in an environment which is multitudinous in its sensory information from hundreds or maybe thousands of other species around it, both plant and animal, the human child is born and grows up in cities, in an environment in which the contact with the mere existence of the non-human is systematically denied.

I believe that much of the stress and the pathological results of that stress, as in dominance hierarchy and striving and competition and all that, arise from sensory undernutrition. Therefore, obviously, the sermon ends with, let children have, at the very least pets, at the very least contact with something, especially at the critical prepubescent time.

We hear that people in the downtown core are suffering from sensory overload, simply because the neon signs flash, and the horns hoot, and so forth. But what we are actually seeing is such a pathetically hair's breadth wide part of the spectrum of potential sensory input that we're not overloaded, we're undernourished to the point of desperation.

David Cayley
Another route back to the roots of our ecological crisis has been mapped by Paul Shepard. He's a friend of John Livingston's and a teacher at Pitzer College in Claremont, California. Shepard's also an intellectual maverick who's dipped into aesthetics, biology, ecology and anthropology in his search for a reconciliation with nature. This search goes back to his days as a graduate student, when Shepard was studying the history of art. His idea then was that we'd stop treating nature as a mere resource if we could only appreciate its beauty. Then he noticed that fashions in aesthetics change and that wild nature has not always been considered beautiful. Paul Shepard went back to the drawing board.

Paul Shepard
I dumped my art books out the window, so to speak, and went back to the library to read anthropology in the hope of finding a sounder base from which to begin my own work. At that time, there was a lot of new work coming in from the study of tribal peoples. For the first time we had good, solid studies in the field of living, foraging, which is to say hunting and gathering, non-agricultural people, which gave us a new sense of the value of the views of such people without trapping us in the old notion of the romanticism of "noble savages."

Well, the thing that astonished me most was not that there was a good, harmonious, ecological relationship among such people and their surroundings, but that their intellectuality about animals was so sophisticated and that their culture contained such erudition. I remember the amazement with which I noticed that Levi-Strauss spoke of Australian aborigines as intellectual snobs after spending an evening with them talking about their mythology. I was deeply moved by this notion that we're not dealing with people who have merely vague and general notions of their world, but highly evolved stories in which a profound heritage of observation and thought made connections to the natural world, largely in analogies between the way in which they organized society and the way the natural systems worked around them. So that's where I then took a whole new 90-degree turn in my thinking.

David Cayley
You can appreciate Paul Shepard's sense of the richness of the hunting and gathering way of life if you think back to something John Livingston said earlier. He was talking about sensory deprivation in modern societies and said we only attend to a tiny fraction of what our senses can potentially tell us. But hunters and gatherers had to maintain a much more universal awareness. Food might appear anywhere, any time, and they could only survive by constantly reading their world for meaningful signs. This ability to "read" their natural world is what Shepard means by the intellectuality of tribal peoples. He thinks civilized peoples lost this ability when they invented agriculture and began to domesticate animals.

Paul Shepard
If I had to pick a moment in human time when what the Christians called "the fall" took place, I think it would have to be with that subversion of ecosystems where we took other species into captivity, into slavery. We talk about animals as friends, or pets, or companions in the same way we euphemistically spoke of human slaves 100 years ago. Taking a substantial number of wild creatures and turning them genetically into monsters that were totally dependent on their human caretakers had some extraordinary effects, I think, on our own psychology, on our way of perceiving the world.

What it did was to create a world around us over which we had control, and which was dependent on our management and manipulation of it. Now we have what? A thousand generations of human beings who've grown up in that world in which the most important creatures are extraordinary blobs — and I can't help using this word because what we do to a creature when we modify it in domestication is to deprive it of all its subtlety, of the richness of behaviour and form that its natural ancestors had. The poor creature has to live in a barren world compared to the environment of its ancestral forms.

So we've struck from these forms that which made them most beautiful and challenging and interesting, and that's our idea of nature. That nature then becomes brutish, it becomes simple, it becomes something we must manage. You can call it stewardship, which is another euphemism we hear occasionally. We are to take responsibility for the world, which merely says that we're on the way to either making slaves of the rest of creation or bringing it to extinction. So that's when the fall took place, if there was such a thing. It was when we left a world to which we can never return, but which nonetheless may give us some important insights and clues as to who we are, when we left that world and became settled people with a very impoverished fauna and flora around us.

David Cayley
Paul Shepard believes that the forager's way of life can still give us important insights into who we are, and possibly show us a way out of our estrangement from nature. The key for him lies in a particular pattern of childhood experiences which our own children tend to lack.

Paul Shepard
Tribal peoples reared their children in the most extraordinary way: by the time the child was initiated into adult society, the person's own sense of living in a world that was good with a great deal of respect for non-human beings struck me deeply. We would perhaps call their attitude a kind of humility about the natural world. Being a biologist, my own bias was to speculate that hundreds, perhaps thousands, of generations of living tribal existence had built into our own genetic make-up certain expectations about what life would be like for us as infants, as children, and as adolescents. That the body and the mind had a programme given by our own biological heritage of assumptions about the way in which we would be dealt with by parents, by peers, by elders and so on, that involves certain critical period experiences.

What seemed to me to be common among this greatly diverse group of foraging peoples was certain kinds of experiences in infancy and childhood that one might call bondings, not only to mother and parents and people, but to the landscape and to the natural world. And then certain experiences that involved a religious perspective, that involved initiatory ceremonies and experiences that the adolescent went through, that moved the individual to a more mature perspective on himself or herself in the world as a whole.

David Cayley
Paul Shepard has laid out this theory at length in a book called *Nature and Madness*. He draws support for his ideas from the vast psychoanalytic literature on child development. But he differs from most psychoanalytic thinkers in one crucial respect. He thinks they've overlooked the importance of our bond with nature.

Paul Shepard
The difficulty is that in spite of the care with which we have been studying children, the people who've been doing it have not, by and large, been very interested in nature or the non-human part of the world. We have this tremendous amount of information, but the interpretations usually neglect even the possibility of some special kind of experience with the non-human. Let me give you an example of what I mean. In the last ten years, studies of infant-mother bonding have led to a whole new sense of the importance of the naturalness of the nursing mother and the importance of contact, sound, smell of the mother, taste of the milk, and the amount of time that the infant spends with its mother. All these elements seem to be essential for the child to come into a world with confidence that there is a protective, informing partner who is nourishing and who is beautiful and who is there.

What we then go on to do, however, is to assume that what the young child does with this is to grow largely into a totally human context. And yet, we have all this information on children in places and in play that suggests what they are in fact doing is transferring the features of that infant-mother bond to some kind of larger matrix. That is, in the third to tenth year of childhood, a kind of attachment to place or places is apparent in the everyday lives of children, that could perhaps be called imprinting. A place is somehow imprinted after the model of the mother as protective, as interesting, as communicating, as nourishing and as infinitely challenging in terms of the bonds that bind the child to it.

Now if there is anything to such a hypothesis, it implies that what the child needs in that period of its life is the same kind of opportunity for bonding to place that we now give the infant with his mother. That what occurs in this infancy bonding with the mother and this childhood bonding with place is the creation in the child of the terms which will serve as metaphors, as the images and the poetic language for creating a total universe, a cosmos, when they come to adolescence. How else do we talk about ultimate things? We talk about paradise, we talk about a landscape. It looks like a pastoral garden. All of the ultimate imagery of what we believe to be finally possible and true about the universe requires a language or other signs and images that come back to social and ecological relationships of infancy and childhood.

*** *** ***

Paul Shepard, John Livingston, Neil Evernden and Allan Drengson do not exhaust, or even fully represent, the field of inquiry called deep ecology. Of the four, only Drengson even identifies with the term. But these thinkers do have a basic approach in common. They all want to probe the deceptive inadequacy of the conceptual languages by which we divide ourselves from nature. And, by stressing the quality of our experience of nature, they all locate the dimension in which real change can occur far beyond the horizon of current political discussion of "the environment." Shepard sees the disruption of "bondings" that would relate us securely to a living and richly informative world; Livingstone identifies a kind of belonging that he believes would be possible were societies not enclosed by a metaphysical dome that prevents us from hearing more than the echo of our own voices; Evernden and Drengson both believe that, until we recognize a participating consciousness, and not a shrunken subject over against a denatured environment, environmentalism will continue to reinforce the alienation it wants to overcome. Taken together, they suggest that, far from solving the environmental crisis, most of us have yet to even grasp its nature.

II
Citizens at the Summit

2

Environment, Development and Debt

In June, 1988 the leaders of Japan, Germany, Britain, Italy, France, Canada and the United States, nicknamed the G7 or Group of Seven, assembled in Toronto for their annual economic summit. Their purpose was to compare notes and coordinate economic policy. At the end of their three days of meetings, the leaders issued a final *communiqué*, praising the status quo, prescribing free trade and free markets as the cure for all economic ills, and counselling a "market-led, growth-oriented strategy" for the poor countries of Asia, Africa and Latin America. They met during the same summer that a big scientific conference was also convened in Toronto to examine anthropogenic changes in the composition of Earth's atmosphere, the same summer that the disastrous synergy of growing deserts, shrinking forests, polluted waters and eroding soils was beginning to make even the people at the top of the planetary pecking order a little edgy. But the G7 had little to say about environmental problems, and nothing to say about their genesis in unequal economic relationships between the rich and the poor countries. The summit's final *communiqué* made a sort of ritual bow to environmental protection and sustainable development, but it seemed more of a political afterthought than a central focus of concern.

Economic summits have been magnets for political protest since their inception in the 1970s, and Toronto continued the tradition. Police removed a tent camp from the grounds of the University of Toronto, where the summit leaders were to banquet. A three-day mock trial indicted the seven leaders for crimes ranging from the Contra war in Nicaragua to the African famine. One hundred forty non-violent protesters were arrested when they challenged police barricades, which had been erected an extravagantly cautious twenty blocks north of the summit site. And, across town at Ryerson Polytechnical Institute, an ambitious counter-conference unfolded. Jointly organized by a coalition of the Canadian left, the Pollution Probe Foundation and a group called T.O.E.S., The Other Economic Summit, which had dogged the G7 since their London summit in 1984, the Citizens' Summit was barely noticed by what one critic called the official summit's "media harem"; but, while these journalists were enjoying the Canadian government's lavish hospitality and complaining about stage-managed news, passionate voices were being raised against the destruction of nature and peoples in the name of economic progress.

Environmentalism is sometimes thought of as a luxury of the rich, but the Citizens' Summit made visible another kind of environmentalism, the environmentalism of those whose very lives depend on an intact ecology. "I come from a part of the world where the prosperity of your part of the world has destroyed our prosperity," Indian activist Vandana Shiva told the conference. "And I bring you a message from the people from whom I have learned ecology, the ordinary peasant women of India."

I spent four rather wild days at the Citizens' Summit, scrambling after delegates from around the world and conducting improvised interviews in a temporary basement studio in one of the conference buildings. From these interviews I composed a four-hour series called "Citizens at the Summit," broadcast in November, 1988. I have included here most of the first two programmes, which focused on the environmental destruction which has followed the forced march of "international development." At the Citizens' Summit there was virtually unanimous agreement that one of the causes of this destruction has been accumulating Third World debt. This debt now stands at well over a trillion dollars. It has driven debtor countries to intensify their exploitation of mines, fields and forests in order to earn the foreign exchange with which to repay loans. In many cases, the projects for which the loans were given in the first place were themselves highly destructive. Barbara Bramble, director of international programmes for the National Wildlife Federation in Washington, D.C., began the programme by talking about these projects.

Barbara Bramble
The most classically easy situation to understand is when a large amount of money is lent to a borrowing government to build a hydroelectric reservoir to produce electricity and/or irrigation water. We've found that some of these dams are being built simply to turn bauxite into aluminum, at a loss to the economy of that country. The world is awash in aluminum and it can be recycled. So, in some cases, there's no need for that electricity in the first place. It isn't being used to help people's lives within the borrowing country, the process of building the dam will displace tribal minorities, destroy valuable pristine tropical forest, and cause immense disease. There are situations where hundreds of thousands of cases of debilitating disease have been caused or at least exacerbated by the creation of a reservoir without proper health measures. The lack of watershed management programmes for a reservoir could result in denuding of the hillsides above it and the erosion of soil into it and eventual siltation of the reservoir, so that in a few decades, it doesn't even produce electricity any more. And yet that country owes the money to pay back the loan that it took out to build this reservoir, so they have all of this harm, displaced people, increased health problems and still owe the money.

David Cayley
So why would anyone do something so manifestly irrational?

Barbara Bramble
There are a lot of reasons. I think the national goals of many countries have been built around producing mammoth projects for nationalistic purposes. It is an important thing for many governments to have their name on a big project. There

is often a short-term employment burst, which is useful to explain to one's people that development is happening. The majority of the long-term jobs, however, in projects like that are usually held by foreigners, certainly by élites and not by the poor. There is an illusion that the electricity is meant for the people, but so often we find that the projects in fact are so much bigger than necessary that they have to almost give away the electricity, subsidize it for industrial use, usually for export and often for products that are in world surplus. It's very difficult to understand, except that there is a mindset that the Western nations have propagated, that these megaprojects are the way to develop, that these are the way to join the modern world. And it's pretty hard for a developing country to resist that argument, it's hard for them to resist money when it's available, and usually the money has been available for specific things. The borrowers know exactly which things are acceptable to borrow money for. They are obviously going to make projects that will fit in with that scheme.

David Cayley
There are literally scores of dams around the world that fit Barbara Bramble's description. A recent case in point is Brazil's Tucurui Dam, the largest hydroelectric dam ever undertaken in a tropical rain forest.

Barbara Bramble
Tucurui has caused the destruction of the lands of several tribal minorities. Those people weren't even told it was going to come to the area until the water actually started rising. I found that very difficult to believe. In fact, I wasn't prepared to see that as one of the crucial problems until I met some of the Indian representatives of tribes in that area who said that indeed, they didn't understand what was happening until the water began to rise around them and they were forced to move in front of the advancing waters and were then found to be, of course, trespassing on the land of the whites in the area. They were shot at, and picked up by the police when they finally arrived at road heads. The tribe has been completely dispersed and the likelihood of it surviving as a tribe is extremely low.

And I recently saw a report which analyzed the macroeconomic effects for the country of Brazil from having borrowed this money. It's a state-owned company that produces the electricity. They sell it at a subsidized rate to the aluminum industry. And at the end of all the calculations, although the industry itself makes a profit because they have the subsidized electricity, the nation of Brazil is paying to produce this electricity because the world price is such that they cannot sell it at the cost of production. They are actually losing $100 per ton. Nobody in any of the parts of this cycle knows that or feels that, but the economists that finally did all the adding up in terms of the national costs and the national benefits have found a $100 per ton loss. This is extraordinary when you think of how easy it is to recycle aluminum, how little that electricity is needed. All countries producing aluminum are in the same vicious cycle. We're all, including the United States, subsidizing electricity to the aluminum industry because there's no other way to sell it. It's a very high-priced commodity if you include the full price of producing the electricity, and recycling costs in energy terms are something like five per cent of the electricity required to produce new aluminum.

Now the reason someone in Brazil could justify this, and in an odd criminal way, it can be justified, is that the costs which were added up that totalled about $1100 per ton, are in local currency, most of them. The price that Brazil receives for the aluminum when they sell it on the international market is in dollars, and they need those dollars to pay their international debt. So at this point, while it looks quite absurd, in fact that $1,000 per ton is vital foreign exchange that Brazil needs to pay this enormous debt which they have accrued over the last decade.

David Cayley
Other countries are being driven along the same path as Brazil. Most Third World countries are producers of primary goods, and in the 1980s, have seen many commodity prices fall at the same time that high interest rates were doubling their indebtedness. As a result, they have been forced to increase exports and let the land pay the price.

Barbara Bramble
A lot of the export agriculture is not sustainable. The soils are eroding, the capacity of countries to produce some of these

products is diminishing, and minerals are being mined at a huge rate. Those that are involved in forestry are pushing the end of their actual forest production. Malaysia, as I understand it, will be at the end of its forest products export possibilities within a couple of years. They have depended on tropical forest hardwood exports as one of their main products. Costa Rica, in an export boom for cattle and rice, has reduced its tropical forest to the boundaries of their national parks system. There are trees left, but they're not saleable, or in many parts of the country, there simply aren't any trees any longer. It's really quite amazing what has happened within just a few years. The accelerating pace has been exacerbated by the debt, but I shouldn't leave the impression that these policies or these ideas were somehow caused by the debt. In some cases, they were accelerated by it, but the idea of subsidizing export agriculture has been the advice given by the major financial institutions such as the World Bank, the IMF and the major donor countries, for years.

David Cayley
The compound effect of debt and misguided development policy has been an accelerating pace of destruction around the world. The Philippines has reduced its virgin forest from 17 million hectares to just over one million in a little over twenty years. Brazil is literally on fire. At one point this summer, airports had to be closed due to smoke from the 6,000-odd fires ranchers and settlers had set in the rain forest. And this deforestation not only destroys an irreplaceable heritage, it also leads to a whole chain of other environmental consequences. Vandana Shiva has observed the process in the Himalayan region of her native India.

Vandana Shiva
In the topics, you get extremely heavy, extremely seasonal rainfall. It only rains for two or three months for the whole year. And if your watersheds are right, if your forests protect your mountain systems where the rain falls in large amounts, and from where origins of rivers lie, you get perennial streams, despite the seasonal rainfall. And that's why tropical deforestation is critical. The moment the forest goes, your rain runs off instantly, and nothing is stored for the rest of the year.

When it runs off, it carries with it trees and soil. During the monsoons there are very heavy floods, landslides, and a lot of disasters of that kind. During the rest of the year, you have water scarcity and total drying up of springs and streams. That is the central ecological crisis. Once your forest starts shrinking, you don't get organic fertilizer and your productivity starts falling. Your animal stocks start decreasing because you don't have water for them. Once they decrease, again you don't have enough fertilizer or enough animal energy to do your agricultural operations, and farm productivity goes down.

David Cayley
Practically everyone has an image of the terrible floods in India, but one tends to assume that it has always been that way. In fact, you're saying these floods are the result of deforestation.

Vandana Shiva
Yes. In my lifetime, I have seen a valley where you could walk ten feet and find a spring, and you'd find water all year round and never any flooding. Now, because mining has started in the catchment of some of our streams, I have seen floods of a kind we've never known before; where entire bridges are washed away. Trains had to be cancelled about two monsoons back, but during the summer there was an absolute scarcity of water. I feel very sad, because when people talk of famine in Ethiopia, they think it's some kind of natural disaster and they think that somehow people there were too ignorant to manage their resources. I've seen in my valley, a rich valley, within a decade turn from rich streams supporting very prosperous agriculture into dry river beds. And when you see a river dry up in your own short lifetime, you realize how other areas that must have sustained themselves have gone dry in the last few years because of a mismanagement of resources called economic progress which has been basically a licence to rape the earth.

David Cayley
In Africa, the story of debt and destruction is repeated. There, the main symptom is the food crisis. Shimwaayi Muntemba

heads the Nairobi-based Environment Liaison Centre, which coordinates the activities of non-governmental environmental organizations around the world. She says that the food crisis and the debt crisis are intimately related.

Shimwaayi Muntemba
The African crisis in my own country, Zambia, has led governments to look for solutions to the food crisis. At the same time, they are looking for solutions to the debt burden, and they are under greater pressure from the debt burden than from the masses who are going hungry. In order to do this, the government has been encouraging large-scale production. Foreign companies have come in, and this has meant land has to be found for these companies, so peasants have had to be moved. All this has occurred in the last two years. These companies are going to grow cotton, which is needed for debt-clearing, for creation of foreign exchange. The government is also saying, as national policy, that food production should be in the hands of small-scale farmers so large-scale farmers can concentrate on export crops, but the land is being taken away from small-scale farmers. What are the implications for household food security? I mean, the whole thing is just a mess because of the great burden of this debt-clearing and the food crisis coming at the same time. We go through periods where sometimes you see a ray of hope, but then that ray of hope goes away.

David Cayley
Shimwaayi Muntemba does not argue that the debt crisis is the sole cause of the food crisis. She sees unequal distribution of land as an equally important cause and recognizes that land reform is the key to food security. The problem is that the debt crisis reinforces the priority given to export agriculture and this makes land reform, which is already politically difficult, practically impossible.

Shimwaayi Muntemba
Land reforms are very critical and in favour of small estates, because you find the developments that have been going on in the twentieth century have marginalized so many small-scale producers. In addition to this, you find that land develop-

ments have resulted in people being pushed to fragile ecosystems. For Africa, we have to be serious when we talk about fragile ecosystems because only 18 or 19 per cent of Africa's land is free from inherent limitations. If you put many people onto fragile ecosystems and they are people that cannot be helped by technology, because they are too poor to adopt more efficient technologies, what do you expect? Of course, the ecosystems will come under even greater pressure.

The problem is that these huge tracts are locked up. I have a friend who commands 17,000 acres. In the same district, there are many people who only command two acres. One of the peasants I interviewed last summer said, "Look, we know the value of crop rotation, but how can I practice crop rotation on two acres? You know," he says, "this is the fifth time I have moved." Each time he moves, he expects to get a greater portion. He said, "I started with five acres, but my nephew comes along, he doesn't have one acre," so he had to cut him a small piece. And so it goes.

All these national pressures are great, yet change has to start with reconstructing the way we relate to land. There is enormous tension and contradiction, and enormous heart searching and pain on the side of leaders, because they know that marginalizing the farmer is not going to bring lasting solutions. On the other hand, they need cash, they need foreign exchange to clear the debts, at a time when prices of these commodities have been dropping. It's really quite a Catch-22 situation.

David Cayley
Behind the debt crisis and its destructive consequences is the international monetary system, a system in which the Third World has little or no say. For example, their indebtedness can literally double as a result of a decision made in the United States to raise interest rates, and the buying power of their currencies can be yanked around by changes in international exchange rates, or periodically devalued by edict of the International Monetary Fund. Arjun Makhijani is with the Institute for Energy and Environmental Research in Washington, D.C. He believes that the international monetary system lies at the root of many of the Third World's economic and environmental problems.

Arjun Makhijani
The U.S. dollar, being the medium of international trade and the currency in which all or many or most international transactions are denominated and most loans are taken out, gives an extremely important leverage to the United States government to manipulate the world economy for domestic purposes. And this has been occurring since the Bretton Woods agreement in 1944. When the system was set up, the United States said we will guarantee this dollar with gold. You can all hold dollars and anytime you want gold, we will give you one ounce for $35. In 1971, in what must be the biggest real debt repudiation in history, Nixon said, essentially, forget the debt. You are all holding these hundreds of billions of dollars. We promised you yesterday we were going to give you a real commodity called gold for it. Today, I am telling you, no deal. I am not going to do it and what are you going to do about it? It turned out the world couldn't do anything about it because there was no alternative medium of international exchange that could be used as a substitute for money in international transactions.

David Cayley
The United States repudiated its obligations because it simply could no longer afford them. It had expanded its money supply to pay for the Vietnam war, and, as a result, there were vastly more dollars in the world than it could afford to redeem for gold. This left the international monetary system without an anchor and, therefore, subject to a series of manipulations that would have disastrous consequences for the Third World. Arjun Makhijani believes that this happened in several stages. The first was the Vietnam-era spending spree, which led to the repudiation of the gold standard. The second was the oil shock of the early 1970s.

Arjun Makhijani
In 1973, oil prices went up and the United States responded by inflating the domestic economy. And, of course, when a large number of dollars were printed, other countries followed suit so exchange rates would stay more or less stable between the currencies. World inflation followed. World inflation means all the goods that were denominated in hard currencies

were increasing very rapidly in prices, and wages were more or less keeping step in these countries. However, wages were not keeping step in the Third World, so the real prices of commodities sold by the Third World were going down all the time. And because the Third World decided to borrow money to pay for these higher prices, a lot of Third World countries became quite indebted.

In looking at the pattern of debt, we were initially told that oil-importing Third World countries were becoming indebted because OPEC had raised the price of oil, but the pattern of debt did not follow this at all. Although oil prices had been a factor in some countries like Brazil, among the most seriously indebted countries were a large number of oil-exporting countries. In fact, in Latin America, there were more oil-exporting countries with serious debt problems than oil-importing countries. Mexico is a prime example, and Ecuador, Peru, Colombia, and Venezuela are other examples. They are all oil-exporting countries. Brazil, and to some extent Chile, are the only large oil-importing countries that are seriously indebted.

David Cayley
It's Arjun Makhijani's contention that the United States and other Western countries escaped the impact of higher oil prices by passing the cost on to the rest of the world in the form of inflation. This increased the cost of imported manufactured goods in the Third World without a comparable increase in the price of their exports. The result was that the current account imbalances that the oil shock should have produced in the West turned up instead in the Third World.

Arjun Makhijani
If you look at the increase in oil prices that the West had to pay for oil imports — forget the Third World for a minute — almost 75 or 80 per cent of the oil was being imported by the United States, Japan, France, Germany, et cetera. You say, well, how much did their oil bill go up cumulatively between 1973 and 1982? In 1973 the price was four dollars per barrel; in 1982 it was thirty dollars. If you add up all the dollars, it's about one and a half trillion dollars. So you should see some balance of payments deficit in these countries to pay for this oil price

increase, but you don't. There was a cumulative balance of payments surplus in these countries in that whole period and deficits only in two years — 1974 and 1980. But this was because oil was essentially paid for by inflating the world economy and reducing the value of the dollar and increasing the value of the exports, and making everybody else pay for the increased price of oil.

David Cayley
The final stage in this process of growing indebtedness was the rise in American interest rates, which began at the end of the 1970s. This policy was the direct consequence of the inflation with which the United States had paid for its oil bill, and it would have equally catastrophic consequences for the Third World.

Arjun Makhijani
By 1979, the second round of this inflation had been started and the system was ready to collapse. Confidence in the dollar had collapsed. Other countries, particularly West Germany, were not willing to follow U.S. inflationary cycles, and they began selling dollars. And at that time, Mr. Volker was appointed by President Carter to stop the fall in the value of the dollar. So this third round of increasing the value of the dollar by keeping up high interest rates started in late 1979. By 1982, for instance, the French franc, which was four francs or so to the dollar in the late 1970s, was nine francs to the dollar, and the situation with many other currencies was similar.

What this did is vastly increase the purchasing power of the dollar and also vastly increase the number of resources, both real resources and amount of money that people had to commit to repay those debts that they had taken. So debts were taken in one circumstance. Overnight, the circumstances in which those debts had to be paid back and the real resources which had to be committed to those debts were changed dramatically, and the debt doubled. Essentially people went to more borrowing and the debt doubled in three or four years. In 1982, the crisis broke upon us. In 1982, the debt was 500 billion, and as a sign of mismanagement of the crisis and essentially the wrong approach to solving it, after six years of saying they are solving the debt crisis, the Third World debt today is one trillion dollars, double what it was in 1982.

David Cayley
The debt has been a driving force in environmental destruction in the Third World, but it certainly hasn't been the only force. The development strategies promoted by aid agencies and international banks were crucially flawed before the debt ever became a factor, and purely domestic forces within the Third World have also played a role. The incredible pace of rain forest destruction in Brazil, for example, has been largely due to the movement of population from southern Brazil to the country's northern frontier, the vast forests of Amazonas.

Richard Norgaard is an environmentalist turned economist who teaches at the University of California at Berkeley. He's studied the economic forces within Brazilian society that have driven settlers to the north.

Richard Norgaard
There's been a long, long history of tremendous income disparity in Brazil, and years and years of pressure for land reform. We have the sort of situation where 60 per cent of the land is owned by 5 per cent of the people. If you redistributed that land, there would be much less poverty and much more opportunity in Brazil. But for political reasons, redistribution would prove very difficult. And so the Brazilian government looked to the Amazon in the north and said, well, there's all this land. We'll open up the land without people to the people without land in the south, and this will relieve the income disparity problems and we'll have a society that we hope will hang together much better.

There are two other major pressures that drove settlers northward. The government subsidizes commercial agriculture. Tax incentives were provided, low interest loans were provided, but the basic problem was that these subsidies were more advantageous to the rich than to the poor. So for twenty-five or thirty years, there's been an incentive for rich people to buy agricultural land from successful, small, rather poor farmers in the south. Subsidies basically gave the successful, small, low-income farmer a tremendous incentive to sell to a rich person, or to a corporation. These people could sell their land for $25,000, and head to the north with their money. In the Amazon, they imagined they could live for five years, establish a farm, get things set up, be a bigger success, and have four or five times as much land. So the subsidies, the structure of agricultural policy in Brazil drove successful farmers away from lands that they should be continuing to practice on.

The third major cause is inflation. Any country that has tremendous inflation has people with money trying to find assets that are not going to be affected by inflation. So you try to put your money into real assets, and there's nothing more real than land. As people try to put their money into land, the price of land goes up, and that makes putting money into land look even better. Land speculation occurs, and people see that putting money into land not only protects my assets from inflation, but that land values also rise faster than inflation. We have situations of 80 to 200 per cent inflation in Brazil and land values increasing by 100 to 300 per cent per year. There's been a tremendous rush to grab land due to inflation.

David Cayley
For the Brazilian government, its northern frontier provided a way of giving people opportunities without actually challenging the gross inequalities within the society. The land without people wasn't really without people, and most of the land wasn't really suitable for agriculture. But the government still went ahead with a number of ambitious colonization schemes. They proved disastrous.

Richard Norgaard
The Brazilian government thought they could design exactly how people ought to live or would be most successful in an ecosystem that Brazilian planners in Brasilia knew nothing about, and they literally set up communities of fixed sizes at particular points along the road. These were communities of twenty to forty families, and everyone was expected to walk to their land from these small centres. The idea of having the small centres was that you could then provide social services. You could pick all the children up and take them to a school five miles down the road, you could have an extension agent come through and talk to all the farmers on Sunday afternoon,

you could provide credit to the farmers much more easily because they were concentrated. The only difficulty was that some of the farmers would have to walk ten kilometres each way, each day to their land, and they quickly grew tired of that, especially in the rainy season when they were knee deep in mud along these roads. They finally moved to their properties, taking their children with them. It became much more difficult, if not impossible, for the children to go to school, and the plan failed. There was a tremendous amount of planning in some projects, but it was false planning, it was disaster, and it wasn't adaptive. The planners never learned. They recommended a specific variety of rice, and that variety turned out to be wrong. It failed the first year. The next year, that same variety was required. You could not get an agricultural loan unless you bought that variety of rice. The plans were that detailed and that wrong.

David Cayley
At the same time as these colonization schemes were unravelling, large-scale cattle ranching, heavily subsidized by the government, was also developing in the north. And all of this was taking place in a "wild west" atmosphere in which ranchers, peasants and native people were often in a virtual state of war. The tragedy was that most of the lands of Amazonas were completely unsuitable for either ranching or agriculture, and this just added to the destruction. Jose Lutzenberger, a Brazilian agronomist, comments on the soils in the settlement area.

Jose Lutzenberger
Some of the rain forest is on sandy soil, some is on extremely degraded clays, and some, where you have rock outcroppings, is even on good clay, but most of it is on soils that are unsuitable for agriculture, at least for sustainable agriculture. But the official settlement schemes are not taking this into account. People cut down the forest everywhere. You only have to look at the maps of the government colonization agency. A complicated landscape with hills and swamps and plateaus and meandering rivers is cut up in parallel and crossing straight lines, like a chessboard. I was once present at a ceremony where settlers received their lots. A few days later, after seeing

their land, many came back desperate. "Oh, I got a piece of rock only." Another one got a piece of high plateau, a cliff and a piece of swamp at the bottom. One settler would have no access to water. The next one had his lot cutting across five meanders of the same river, so he had to build five bridges if he wanted to work that land. Even the land is cut up in a way that simply does not take into consideration what exists. It's all done abstractly on a map, so some people happen to get good pieces of land, but most get very bad.

David Cayley
Let's say you are amongst the most, who have very bad. What will happen to you?

Jose Lutzenberger
What has already happened is that most of them have abandoned their land. If you look at the first settlements in the state of Rondonia, most of the settlers have given up, and some, as we show in one of our films, have already resettled for the fourth time. They give up and go to another place, give up again, and keep on going. And this is why the process is so destructive, because people can only survive by causing ever more destruction.

David Cayley
Environmental destruction in the Third World is now giving rise to resistance everywhere. In Malaysian Borneo, forty-three members of the Penan tribe are awaiting trial for blockading logging roads. In Brazil, native people, rubber tappers and other forest dwellers are defending their way of life. And in Africa, there are now over 300 non-governmental organizations involved in environmental protection. One of the inspirations for this world-wide resistance movement has been India's Chipko, which means literally "embrace the tree." This movement of opposition to commercial logging began in the early 1970s in the Garhwal Himalayan region of north-eastern India, but it had deep roots in Indian society. Forests have always played a central role in the culture, economy and religion in India, and under British rule, there was already widespread popular opposition to the expropriation of forests. Vandana Shiva comes from the Garhwal district, where

Chipko began, and she has been deeply involved in the movement.

Vandana Shiva
There is often an assumption that environmental concern comes only to those who are affluent enough to be in the Western industrialized world, or to be the élite of the Third World. That it's something that comes after you have become very well-to-do. The second assumption, which is linked with this, is that somehow the poor of the world cause environmental destruction and have to be taught about environmental protection. Long before the U.N. Stockholm conference on the environment happened in 1972, where the government agencies of the world decided to pledge themselves to environmental protection, the women of Garhwal had been fighting to protect their forests, not as an exotic, esoteric kind of conservationism, but as the very basis for their lives. For them, the conservation of trees is not in opposition to their livelihood, and ecology doesn't work against economics. For them, the basis of their economics, the basis of their production of food, the basis of their livestock management is all the forest. The forest provides the fodder, the forest provides the fertilizer, the forest provides the water which makes agriculture possible.

As deforestation expanded, largely for commercial reasons, agriculture, which women take care of, started to become more critical. Where women used to draw water five minutes away and used to get fodder ten minutes away, they were now having to walk for five or six hours a day to collect the same needs. At some point, the women very clearly said this can't go on. When the next contract was given for logging, the women started embracing the trees and saying, "You are killing us anyway, so you might as well kill us while we protect the basis of our lives, while we protect our mothers" — that's what they call the forest — and actually drove the loggers away.

Most places, people didn't even have to embrace the trees, but in 1977, police were brought in to protect the logging company. The women literally clung to the trees and had to be dragged away, but they came back the next day. The foresters tried to tell these village women that they were blocking a national activity that generated revenue. And the women, they came with lanterns in the daylight, and they said, "We bring you light. We bring this lantern in the daylight to show you light because you might have all the degrees and we might not have any education, but we do have the education that our mothers gave us. According to that education, these forests are sacred. These forests are the basis of our lives, and these forests do not produce resin and timber and revenues. The real product of these forests is water and soil, and we are protecting that economic base."

In different pockets at different times, village communities protected their forests, and they did it so systematically for such a long period — over a decade — that by 1981, the government had to put a ban on commercial logging in the Himalayas.

David Cayley
This logging ban and the forest conservation act which Mrs. Gandhi's government enacted at the same time were great victories for India's environmentalists, but they were just the beginning of a continuing struggle. The Forest Conservation Act is now under heavy counterattack by logging interests, and other destructive developments have already been approved.

Vandana Shiva
There is a particular case of a dam on which I am now working, an evaluation. It's a World Bank-financed dam on the Subarnarekha River in Bihar. I discovered that for the past ten years, people have been protesting against the dam. In 1978, six people were killed at a protest. In 1979, one more person was killed. In 1982, the leader, the chief of that tribe was murdered by the police, who called it a death encounter. And in August last year, another tribal chief ...

David Cayley
When you say they called it a "death encounter," you mean they claimed it was self-defense?

Vandana Shiva
Yes, they use that phrase. Dissenters are called "extremists," and all extremists die in "encounters" with the police. That's

typical police jargon. I went to this area and it suddenly hit me that here's an area about which I am concerned, and I do know about the river that flows through this region. But even I, a concerned ecologist, did not know that eight or nine people had been killed by the police for resistance against the dam. Every time the government has tried to squash the protest by killing somebody, the people have banded together more to fight against the dam project.

David Cayley
What's the purpose of the dam?

Vandana Shiva
The government says it's for irrigation. It's fascinating that when I finally went to the site to see how it would help, I found these people have their own indigenous irrigation system. What the government is going to do is destroy this indigenous system and have large-scale canals, which, in an up and down topography with an extremely hard rock area underneath, are sure to cause very severe waterlogging problems. The project reports admit it. But the reason they're going ahead with the dam is the unwritten line in the proposal, that ultimately the water is meant for an industrial township for a monopoly house that manufactures steel in India. It's really for a steel factory that is expanding and needs more water. And so they're building a dam for an irrigation system, which is actually an industrial system, in which they are borrowing from the World Bank at public cost to subsidize a private industry.

David Cayley
The opposition to the Subarnarekha dam is characteristic of a new environmental movement now appearing around the world. This movement is the very opposite of the common stereotype of environmentalism as a middle-class avocation, something to do when your belly is full and your economic needs have already been met. It's a movement of people who have never divided ecology from economics, people whose environment is their economy. These people are making last-ditch defences of their life places all over the world, and in the process, through movements like Chipko, they are recreating their traditions.

Vandana Shiva
Right now, Chipko has become, to me, India's ecological philosophy, an expression of an Indian view of an environmental movement. The difference about it is that it's the poorest of the poor in India who are leading the ecological movement. They don't have to be taught that nature must be protected. It's in their own categories of thought. It's in the context of environment not being a package of resources that you must conserve so that economic processes can somehow sustain themselves a little longer. Nature is Prakriti, and Prakriti is basically the life force of earth. What you're protecting is the life force that gave you life, and that's the language in which people talk, in which the movement has been born, and it's responding to the needs of the people.

That's why at an international forum, when people talk of how the Third World can't afford environmental protection, I always think of how it's the poor of the Third World who are enforcing environmental protection on governments who would like to destroy natural resources for a quick buck, for a short term in office.

Today, to be an environmentalist in India means standing with the tribals and backing them up, even if the police might be brought in to shoot at the protest side. The personal costs to keep the fight going are increasing, and that's the point at which real commitment shows up.

*** *** ***

The environmental movement which Vandana Shiva is describing is paralleled in Canada only in those few remaining places where people still derive their livelihood, and the meaning of their existence, directly from the land. It is a movement which has grown in lock-step with development-induced displacement of people and destruction of life places. The Citizens' Summit manifested the existence of this new environmentalism in a compelling way. The big media, unfortunately, were looking the other way, transfixed by the smoke and mirrors of the official summit. The fact remains that, throughout the world, traditional societies are standing at the last ditch, fighting to defend a way of life in which ecology remains integral against the development that is destroying it.

3

"You Have to Keep Swimming"

The second programme of "Citizens at the Summit" continued the analysis of international development begun in the first programme. For forty years, the term *international development* has stood for the idea that the kind of economic development that has shaped the modern West could somehow be reproduced in the rest of the world. Its results have been paradoxical. Tanzania, for example, is Africa's biggest aid recipient. It receives an annual subvention which is greater than either its tax revenues or its export earnings. But Tanzania is also an economic basket case, a country that has devastated its agriculture through forced collectivization and central planning. There's a connection. In Tanzania and elsewhere, foreign aid has insulated governments from their people and allowed unrepresentative élites to create plans and projects which the majority of their citizens never asked for and didn't want. It began with the idea that there was a kind of formula for development that could be abstracted from culture and history and transferred from one type of society to another. Foreign aid financed the process by providing a source of money for which neither the donor nor the recipient government was actually accountable. Traditional economies and the ecosystems with which they meshed were destroyed, the people displaced. Much of this damage has been done by so-called mega-projects, the classic case being the oversized hydroelectric dam, as Barbara Bramble pointed out in the first programme. In the early 1980s India's Planning Commission undertook a comprehensive analysis of hydro development in that country. They discovered that costs have often exceeded benefits, sometimes by a ratio of two to one. The human tragedies have been incalculable.

Ill-considered hydroelectric dams are but one aspect of a larger problem, a problem that was already widely evident by the early 1980s. Caught in the downward spiral of expanding populations, falling commodity prices, increasing debt and misguided development policies, many Third World countries were literally destroying their environments. Landless peasants and unemployed workers were pushed on to marginal lands. Hillsides were left bare and eroding. Forests fell to the advance of settlers and loggers. In 1983, the United Nations responded by creating the World Commission on Environment and Development, later known by the name of its chairman, Norwegian prime minister, Gro Harlem Brundtland. The twenty-three commissioners came from all over the world and held hearings all over the world.

The Brundtland Commission's report, *Our Common Future*, was published in early 1987. It called for a new era of economic growth, which would be based on the understanding that all economic activity depends on our ability to sustain and renew the natural environment. Their secretary, Chip Lindner, says that the report was based on three main principles.

Chip Lindner

The first is that environment and economics are two different sides of the same coin, that no longer can we look at the environment as unrelated to economics, or vice versa, because the two are inextricably linked. Confirming and getting that concept accepted in a much broader political circle has been one of the commission's major contributions. It is something that many have known for some time, but now has become far more accepted and is being much more broadly acted upon.

The second principle is that poverty has become so widespread, so pervasive around the world, and is causing such fundamental impacts on the maintenance of the basic resource base, that unless poverty can be turned around, there is very little hope of a sustainable future for any of us. And there is little sign that that's happening. Therefore, the commission came to the conclusion that there had to be a new era of economic growth, that poverty could only be turned around through new growth, but the growth had to be entirely different. It had to be environmentally, economically and socially sustainable, to have a different content, to be concentrated on clean technology, and a total and complete recycling of waste, and to entail fundamental revisions to the disparities and inequities that exist in the international economic relations system.

The third principal element that runs through the report is the question of people's participation and the necessity of involving the beneficiaries, or what are more often the victims of development, in planning for future development. Without involving the people who will ultimately hold the responsibility of maintaining the development system in place, you have little chance of the project or the development process succeeding.

David Cayley
The Brundtland Commission never questions the value of economic growth. The commissioners are convinced that poverty and environmental destruction form a vicious circle, each causing the other in an endless downward spiral. Therefore, they see growth as the only way of protecting the environment from the destructive effects of poverty. Consequently, they neither challenge the idea of development itself, nor examine what was once the central thought of environmentalism, that growth is unsustainable by definition. The commission's politics, in other words, tend to be reformist, favouring an ecologically managed capitalism. But though the report is far from revolutionary, its programme of reform is extremely ambitious and far from being assured of success. It calls for changes that threaten real state and corporate interests, changes ranging from land reform, to forest conservation, to a revamped international trading system. The report also calls for popular participation in development decisions, and that is what appeals to Pat Adams, executive director of Probe International. Probe International is a Toronto-based organiza-

tion that monitors the effects of Canadian aid and trade policies on the countries of the Third World.

Pat Adams
The Brundtland Commission report is the product of a diplomatic report-writing exercise. As a result it's written for all people, but I do think it actually has an important underlying theme that the world should latch on to. And that is that at the very basis of sustainable development is democracy. The way you come up with constructive development projects that people want and that don't destroy the environment is by asking for participation, and by giving people the tools to refuse projects that undermine their right to a clean and sustainable environment.

And that's what Mrs. Brundtland talked about. She talked about democracy, and she even recommended that referendums be held for some of these mega-projects. When we see environmental destruction, I often think it's because the wrong people are making decisions. It's because the people who don't have to live with the consequences of those decisions are the ones who are making them.

David Cayley
Right, but for the diplomatic reasons you alluded to earlier, the report doesn't actually make clear who those people are, except in the most veiled terms.

Pat Adams
That's right.

David Cayley
So can we see dangers? Is the Brundtland report co-optable?

Pat Adams
Yes, I'm sure it is. In fact, we're seeing that right now. The term *sustainable development* is just bandied about by governments all over the place and governments who are not respecting many of the recommendations of the commission at all.

David Cayley
Is Canada possibly an example?

Pat Adams
Oh, absolutely. At virtually the same time that CIDA, our national aid agency was announcing that it was going to be the first Canadian government department to follow the recommendations of the Brundtland Commission report, CIDA was negotiating a secret deal with the Department of the Environment to exempt itself from the conditions of our environmental assessment review process, which calls for release of environmental assessments to the public and for a public review process. They cooked up this very neat little deal with the Department of Environment, saying in the case of our projects, we don't want to follow these particular rules. And the Department of Environment agreed to it. Well, that's outrageous.

David Cayley
Co-optation is clearly one danger the Brundtland programme faces. There are other challenges as well. The main one is the structure of the international economy, a structure rooted in four hundred years of colonial and neocolonial exploitation. Without changes in this structure, it's difficult to see how the commission's plans to save the environment by improving the economy can be realized. Colonialism shaped the economies of most Third World countries. They became exporters of raw materials and importers of manufactured goods from capitalist centres of the north. This left them completely at the mercy of international forces over which they had neither control nor even much influence, and it led to a persistent tendency to over-exploit the environment in response to poor terms of trade.

One sign of this external domination of Third World economies is the power of the International Monetary Fund, or IMF. Over the last ten years, the cumulative debt of the countries of the Third World has steadily increased and now exceeds one trillion dollars. The IMF has responded by imposing what it calls "structural adjustment." Countries in balance of payments difficulties are given temporary credit in exchange for fiscal reforms. Arjun Makhijani describes them.

Arjun Makhijani
Principally, two or three things are prescribed in IMF medicine. A serious reduction in government spending, often tight-

ening of wages and elimination of subsidies generally to the poor, for instance, perhaps in the form of keeping transportation prices affordable, or wheat prices or bread prices affordable, which involves government spending. And the other is to devalue the currency to encourage exports and discourage imports. So that's the policy. It's a very simple formula that is uniformly applied in all situations. In fact, it was invented so a single IMF bureaucrat could quickly do the calculations on the back of an envelope and make his prescription. That was part of the rationale for this kind of formula. Besides, it seemed to be theoretically good.

David Cayley
Arjun Makhijani is with the Institute for Energy and Environmental Research in Washington, D.C. He thinks that the IMF medicine is making the disease worse, and he has evidence. In the last six years, the Third World debt has doubled. He thinks that the problem with the IMF formula is that it doesn't take the real nature of Third World economies into account.

Arjun Makhijani
The IMF formula assumes that all countries are the same, that every country has a very similar economic structure. So that if you devalue the currency, the economy of the country is diversified enough and there's enough demand out there for that diversity of products that exports will increase, and that if you devalue the currency, that imports will decrease. Imports generally do decrease. It's a good way of squeezing consumption. That we can see; it has happened. Consumption has been squeezed.

However, on exports it's a much more difficult picture because you have got, say, a copper-producing country, Peru or Chile or Zambia, which produce one or two primary commodities. They're not diversified economies in their export sector. You devalue their currencies and it does not raise the demand for copper a bit. The demand for copper is dictated by very much larger technical and economic considerations, and is really relatively insensitive to price, particularly over the term of one, two or three years. So what you do is aggravate the condition of the country by squeezing imports without really doing anything serious to exports. That's one kind of reality that the IMF formula does not take into account.

The other kind of reality that it doesn't take into account is to assume that problems are internally generated, that people are demanding too much, and importing too much. It completely ignores the fact that the currency of a single country, the United States, is the international reserve currency and the medium of trade, so that if the United States raises its interest rates, for instance, the value of the debt goes up. In addition, the IMF formula provides a tremendous incentive for rich people and élites in the Third World to export capital to the United States because they can earn a lot of money simply by putting it in a bank, and exports of capital by rich people in the Third World and élites have been a principal part of the debt problem.

The third factor that comes in in devaluations is you are effectively reducing people's wages, so you are cheapening the resources that they are producing. You are reducing their price. That is, for one hour of productive labour that was already underpaid yesterday, today you are paying them 20, 30, 40, 50 per cent less. So naturally, if they are selling their own labour at a cheaper price, but they have to buy commodities from the United States at more and more expensive prices, they are going to have to sell more and more just to stand still. And this squeezes imports even more and makes labour even more artificially cheap, and intensifies resource exploitation, environmental degradation, and so forth.

David Cayley
To solve these problems, Arjun Makhijani believes that a complete restructuring of the international monetary system will be necessary. This would involve, at a minimum, a new, truly international medium of exchange to replace the U.S. dollar and a much less discriminatory way of setting exchange rates. Without such a system, current IMF policies will continue to drive Third World countries deeper into recession and debt and make the Brundtland Commission's hopes harder to realize.

The Brundtland Commission's hopes also rest on a quick solution to the Third World debt crisis. Right now, the debt stands at over a trillion dollars, and a growing number of countries are pouring more than half their export earnings into servicing it. This translates into both economic recession and environmental stress. The debt crisis was discussed at the Toronto economic summit last summer, but no binding course of action was proposed. Instead, the Group of Seven leaders agreed to what they nicely termed "a menu approach" to the problem. This leaves each of the seven free to devise an individual response from a menu of options, ranging from outright forgiveness to extended repayment terms. Canada and West Germany have both already forgiven some government-to-government debts.

Yet another obstacle to the Brundtland Commission's dream of sustainable development is presented by the various national and international aid agencies that currently support unsustainable development. One agency that's already come under heavy pressure from environmentalists to change its ways is the U.N.-affiliated World Bank, headquartered in Washington. Barbara Bramble has been deeply involved in these campaigns.

Barbara Bramble
We've been working to reform the World Bank for years, because it has such a great influence on the other funding institutions for development. And the World Bank itself, over the last year, has announced a major series of reforms. They have developed an environmental department within their own structure. They have reorganized environmental offices into their operations departments. They have added environmental personnel — not as many as we would like to see — but people trained in a wide variety of disciplines that they have needed for a long time. They've gone from perhaps five people in environmental positions around the bank up to more than fifty, and they're aiming for one hundred. Their new policies on wild lands conservation and what to do about the forced resettlement of people are models. They are really extraordinary documents. The question is how they will be implemented and that's what we're watching now.

David Cayley
Whether or not they are implemented, there are some ecologists, like India's Vandana Shiva, who think that the Bank's

reforms will remain window dressing. Vandana Shiva works with ecology movements in India and around the world. She thinks that the Bank's very reason for being is problematic, and that makes her wary of superficial reforms.

Vandana Shiva
In things related to politics at any level, you can't ever afford to allow serious issues that touch millions of people in various ways to ever be reduced to a kind of token. The system that creates the destruction must never be allowed to say, look, we're giving you a solution because this one project you made noise about, well, we've fixed it. If your mindset is based on increasing profits, increasing links with global markets, changing all subsistence systems, and all systems that produce livelihood for local people into systems of profit generation for a small minority, then unless that philosophy changes, destruction will continue. Part of our work in the Third World is to show that you might talk of the Narmada Dam and the World Bank might say okay, we'll be a little more careful about its environmental impact, but after the Narmada, there are ten other dams, and beyond the Narmada, there are twenty other water projects. The main thing, these are very fundamental changes. They're as basic as what happened at the time of enlightenment, when entire world systems were redefined. What is the world made of? What is society made of? That's the kind of watershed we are on.

I think one has to debate with the World Bank on individual projects, but what's at stake is a redefinition of extremely basic categories about what different cultures are. Are they backward because they don't do things like us? Are they backward because they don't do science like us? Are they poor because of their own fault or are they poor because we, to get rich and have more, deprive them of what they had? Is nature a block-in-progress or is it the limit within which all human well-being must be defined? These are the kind of very basic issues with which ultimately we must all struggle.

David Cayley
These basic issues are now being addressed all over the world, and in that, there is reason for hope as well as despair. Perhaps the right kind of hope will appear only when we have

despaired. One of the most heartfelt presentations that was made to the Citizens' Summit was by a Brazilian agronomist called Jose Lutzenberger. He talked of the destruction of rainforests and of the extinction of species. He talked of the peoples who have disappeared and the languages that will never be heard in the world again. And he talked with such honest feeling that I wondered how he could stand to live with this terrible knowledge. The next day, I got a chance to talk with him and was surprised to discover a man full of hope. Institutions lag, he says, and continuing destruction is virtually certain, but people are changing.

Jose Lutzenberger
In Brazil, we have a situation that is really unique, I would say, on the whole planet. I know European agriculture, I know American agriculture, and there is nothing like it in the world. To give you one example, we have two important agricultural magazines in Brazil. Together, they print something close to a million issues every month. These magazines — without ever mentioning organic farming, alternative ecological or regenerative farming — promote almost exclusively exactly that, because farmers want it. This has happened because a majority of our agronomists are now looking for alternatives. They want this kind of literature, and the farmers have reached the conclusion, at least a great proportion of them, and this is without any idealism, that pesticides and the conventional forms of agriculture are just too expensive. They're looking for alternatives. Even within the system, there are many, many people who would like to change. When I go back to Brazil, I will be at a meeting in Rio Grande do Sul with more than 200 Brazilian agronomists. The same people who, a few years ago, thought that I was a little crazy are now inviting me to help them change.

David Cayley
After listening to you last night, I wanted to ask you how you could stand it, how you could live with the kind of devastation that you described, feeling as it was evident that you feel. Now I'm getting a different impression. I'm beginning to see that **maybe you have more hope.**

Jose Lutzenberger
I'm fundamentally an optimist and I am more optimistic today than I was eighteen years ago, when I started the environmental fight. Then, I was convinced that if modern industrial society, especially its last degenerative outgrowth, what we call the consumer society, if that lasted another thirty years, that would be the end of life. But by now, we've already seen things changing. They're changing all over the planet and a meeting like this one here is proof of that. World-wide, I find people who are beginning to think differently, and even among technocrats.

Right now, I am working with a big cellulose company, and this is perhaps the greatest satisfaction of my life. For twelve years, we fought a tremendously polluting pulp mill across the river from my home town. They caused tremendous pollution in the river, but today, it is perhaps the cleanest pulp mill in the world. They spent 40 million dollars setting up a treatment plant. They also have perhaps the best forest service I've seen, even though they make enormous eucalyptus monocultures. About 30 per cent of the land they control is nature reserve, with native forests intact and full of fauna. It's the only nature reserve we have in our state that is really protected. Together, we are developing methods for turning the biological sludge from their treatment plant into an organic fertilizer. So, a former enemy has now become a powerful ally.

And I think there is so much we can do in this direction. So much is already happening. I have already worked with other companies and we have shown, especially in the field of sanitation, that what was formerly considered waste is actually raw material. It is not a question of controlling pollution, it is a question of simply not making it, or turning what was supposed to be dirt into merchandise. But there is an awful lot to be done.

Of course, we are heading for catastrophe, that's inevitable, but perhaps we can soften the fall. On the other hand, even if I saw no light at the end of the tunnel, I would still be fighting. Let me use a final image. Suppose you found yourself swimming in a shark-infested sea, 30 miles from the coast. The chances that you would make it are almost nil, but would you stop swimming? You have to keep swimming.

*** *** ***

It is perhaps a sign of the times that in 1990 Jose Lutzenberger was appointed Brazil's Secretary of the Environment by the newly elected government of Collor de Mello. In the capitals of the West, environmentalists maintain steady pressure on development agencies to reform their practices and limit the destructive effects of their projects. In the Third World, grassroots resistance to big development projects grows. But what is at issue, as Arjun Makhijani points out, is not just this or that project, but the entire structure of international economic relations. As long as poor countries remain tied to the treadmill of debt, unequal terms of trade, and the desperate pursuit of foreign exchange earnings, sustainable development will remain a will-o'-the-wisp.

III

From Commons to Catastrophe: The Destruction of the Forests

4

The Last Assault

Twenty-five years ago I lived for two years as a Canadian University Students Overseas volunteer teacher in the little Malaysian state of Sarawak, which occupies the northwestern section of the island of Borneo in the South China Sea. When I arrived in 1966, I found a country remarkably undisturbed by development. There were government schools and a rudimentary road system; but most travel was still by river, and the longhouse peoples of the "interior" seemed to have integrated transistor radios and outboard motors without any fundamental disruption in the continuity of a way of life still based on hunting, fishing, foraging and shifting agriculture.

I left Sarawak in 1968, and eventually lost touch with what was happening there. Then in the summer of 1988, I saw a story in the Toronto *Globe and Mail*, which reported that in the intervening twenty years the country had been so intensively logged that the subsistence of many of the forest peoples had been destroyed. Customary lands had been turned into timber licenses, the forests that once yielded foods, medicine and shelter had fallen, watersheds had been ruined, and the affected people had finally been driven to desperate acts of resistance, ranging from burning bulldozers to blocking logging roads.

I had been following the issue of tropical deforestation, in Brazil and elsewhere; but learning of its impact on people and places I had known at first hand in happier circumstances gave the question a new pathos and a new immediacy for me. What was at stake in Sarawak and elsewhere, it seemed to me, was not just biological diversity but cultural diversity as well. Burning and clear-cutting tropical forests may or may not alter Earth's climate, but it will certainly destroy the peoples whose way of life is so intricately adapted to them. And with them will disappear unique and irreplaceable ways of being human. A handful of radio documentaries wasn't likely to make much difference, but it was what I knew how to do, and so I produced a week-long series, called "From Commons To Catastrophe: The Destruction of the Forests," broadcast in June, 1989. The first programme, which concentrated on tropical deforestation, follows.

David Cayley
Deforestation is as old as civilization itself. The Minoan civilization of Crete was running out of wood for its bronze foundries and palaces by 1700 B.C. In 4th-century Athens, Plato lifted up his eyes to the ruined hills of Attica, where the rainfall, he said, just wasted off the bare earth. There were wood riots in Tudor England, and hardly a forest left standing in 18th-century Connecticut. But always, these essentially local deforestations occurred in the context of a world that still contained vast wildernesses and unexplored frontiers. Today, the last frontier is in sight, and country after country, from Haiti to Nepal, is discovering what it's lost only after it's gone. Forester Kenton Miller, who today works for the World Resources Institute in Washington, has seen the change in his professional life.

Kenton Miller
Wherever I travelled in those early days throughout South America, Africa, Asia, up to Alaska and northern Canada, I always had a sense that there were plenty of forests, and that things were still growing while we were cutting. I've now had the opportunity to go back to places in the upper Amazon where I did college thesis work (that's a while back), to see areas that we crossed by dugout, for days and days on end

without a break, now completely cleared, down to mineral soil, the soil eroding bare, down into the rivers. And the loss of forest — the net loss of forest — they may come back into something green, but it will not be a forest with value either to timber production or to any sense of representing the range of species that used to be there. Things are oversimplifying. The word "impoverishment," getting poor, is now very, very real. I feel a sense of desperateness that the thing is really going in the wrong direction. We are definitely setting up a stage for destruction, not simply losing a few things here and there. The scale of change is out of control.

David Cayley
Concern with this scale of change now focuses on the lands that lie between the Tropic of Cancer and the Tropic of Capricorn — the tropics. It has its ironic side, coming as it does from countries like Canada, that have grown rich by exploiting their forests. But there is more to it than just self-righteousness. Tropical forests are different, different in the strategic role they play in the biosphere as a whole, and also different, says Brazilian agronomist Jose Lutzenberger, in the extraordinary abundance of life forms they contain.

Jose Lutzenberger
The tropical rain forest is the richest living system on the planet, and I'm referring not only to South America but also to Africa, Asia, Indonesia, New Guinea, and so on. An incredible diversity of living beings has been able to evolve. When you look at a European forest, or a North American forest, we have a few dozen species of trees. I don't know how many it is in Canada. In Europe, it's very little. In the rain forest, botanists have told me that they have already described something like three to four thousand species of trees and there are at least 10,000 more to be discovered and described. When you go to orchids, herbs, ferns and so on, and when you look at all the animal world, especially the invertebrates, insects and so on, then it is millions of species that we don't even know yet.

David Cayley
The reason for this striking difference in species diversity between tropical and temperate forests is pretty much what

you'd expect — the weather. Adrian Forsyth is one of a handful of Canadian tropical ecologists, and the author of *Tropical Nature*.

Adrian Forsyth
What makes the tropics so rich are all the different things that an organism can be in the tropics, as opposed to the far north. It's just easier to be a vine, for example, in an Amazon forest, where you don't have freezing and thawing or as much temperature fluctuation within a single day and night as the tropics have during the whole year. It's very easy to evolve a very specialized solution like, say, stretching your body out over half a mile and becoming a liana, whereas when you get to Ontario, you know what happens if you have long, undefended plumbing. It just gets blown to pieces by the spring and fall freezing and thawing.

So it's all the interesting ecological release that goes on there that allows for all kinds of different plant life forms that simply can't exist out of the tropics because of the physical nature of water and proteins. You know, they can do certain things at certain temperature ranges and the tropics are optimal for those. When you allow all these different kinds of plants to exist, then you allow all sorts of different animals to do different things. So again, you can have animals that make their entire life in the treetops, and you just don't have that in the temperate zone. And because every plant then supports about twenty other species on average, for example, different kinds of insects and mites and things that feed on the insects, you get this concatenation effect. Every time you introduce a plant species into a community, you add all kinds of other organisms that can then depend on the plant, and they depend on each other, and so on. It starts to snowball.

David Cayley
Ecosystems are not only richer in the tropics, they're also more highly localized. Biologists call this phenomenon endemism — the occurrence of species in one place and nowhere else.

Adrian Forsyth
It's hard for a lot of people in temperate zones to realize. If you know the trees in Ontario, for example, and you go to Nova

Scotia, hundreds and hundreds of miles away, you see the same maples and oak trees and not much difference in the flora. But if you cover that same amount of distance in the tropics, you've just passed over community after community of different trees. You go just a few miles and you're finding new species. And so, every time you lose a small patch of habitat in areas of the tropics, you're losing species. So you just have to clear a mountain top, for example, and you've lost a number of species that existed only on that mountain top and nowhere else. Whereas, if you take a nice patch of pine forest at Temagami, it's the same pine forest, in a sense, that you're going to get hundreds of miles away.

David Cayley
The limited range of many tropical species is one of the reasons for the high rate of extinctions we are now seeing. Another is the many mutualisms that exist in tropical forests. Organisms co-evolve in dependency on one another, and so they tend to disappear in groups.

Kenton Miller
The damage that can be caused by felling some of the large trees on the smaller ones is subtle. For example, in the South American tropics, we're finding, in the work of John Terborgh and others, that some trees are only pollinated by certain beetles, certain insects, and they live in the crown of the trees only. If we fell the representatives of that species for so many square miles, we also destroy the pollinators of that tree. That tree may never regenerate again, at least not in that area. When you fell those big trees with the crowns that are enormous, crowns that may cover a quarter acre, they pull down a lot with them: the vines that connect them to the ground and to other trees, the plants and animals that live, in many cases, up there. They fall to the ground. They can't simply walk away and find another place. And they're only found maybe in that one tree species, and there's not another tree of that species for another ten or fifteen acres. They evolved with those trees over millions of years and there's a great partnership going on, up in those treetops. That's a much more complicated situation than we had in the northern temperate zone.

David Cayley
The current rate of species loss, according to biologist E. O. Wilson, exceeds anything found in the last 65 million years. If the trend continues, we may see the loss of up to one-quarter of the world's species within the lifetime of today's children. Human beings, in other words, are now altering the biosphere in ways parallelled only by the great geological and cosmic upheavals of the distant past. This certainly raises a harrowing philosophical question about the right of one species to wantonly destroy so many others, but it also raises very practical economic questions. Tropical forests, historically, have been the source of many invaluable agricultural and medicinal plants, from tomatoes and potatoes to quinine and cortisone. Future discoveries depend on there being a forest to find them in.

The forces driving deforestation are complex and vary from region to region. Certainly, the leading cause, world-wide, is the search of landless and unemployed people for more food-producing lands. In the Americas, ranching has been a major cause. In parts of Asia, Africa and the Caribbean, a critical fuel wood shortage, the poor man's energy crisis, has put intolerable pressure on forests. In Southeast Asia, logging has played a prominent role. Often, logging and agricultural colonization work in tandem, with the loggers opening up the forest and building roads, and the settlers following up to finish the job.

The pressures on tropical forests are intense, but they are not necessarily inevitable. The landless are not necessarily landless because there's no land, but because land is monopolized. Brazil is a startling case in point. In Brazil today, less than one per cent of the farms account for nearly half the land under cultivation and government policies have fostered this concentration by favouring large monocultures of export crops, like soya beans, over domestic food production. There is no land shortage. Land redistribution and new agricultural policies could have accommodated all of Brazil's farmers on productive soils. But instead, the military government, which seized power in 1964, chose to divert land pressure into its vast northern forest hinterland by launching what the generals called Operation Amazonia, a package of policies designed to stimulate agricultural and industrial development in the Amazon region. It initiated what will probably be remembered as one of history's greatest ecological follies, the attempt to establish a productive agriculture on lands which one ecologist describes as "a desert with trees."

Jose Lutzenberger
The soils under the rain forest are the poorest soils in the world. It is very misleading when you look at those fantastic forests. Even some scientists like Humboldt in the last century, when they saw that luscious forest, thought that it must be standing on one of the most fertile soils in the world. It's just the opposite. The rain forest survives because it has somehow learned — if we can apply that word — to recycle all its nutrients as fast as possible.

When you look at a forest in a temperate climate like Canada, you have a situation where most of the nutrients of that ecosystem are in the soil, and between two and five per cent is in circulation in the biomass. The tropical rain forest is the opposite. More than 95 per cent of the nutrients are in the biomass, and there's almost nothing in the soil. The dry leaves that rain down from the canopy of the trees are reabsorbed in a few days, at most in two weeks. There's hardly any bacteria in the soil, and the hair roots of the trees come out of the soil and go into the dead leaves. In symbiosis with certain fungi, the nutrients are taken out directly from that leaf, without going into the soil and back up into the canopy. And when you cut down that forest, burn it, then these poor soils are hit by heavy downpours, enormous rainfalls of up to 5,000 millimetres in some places, and year-round temperatures above twenty-five degrees or so, the soils erode and nothing is left. Of course, you don't get a typical desert with the sand dunes, but what you get is poor, unproductive scrub.

David Cayley
Poor, unproductive scrub is precisely what now exists on large tracts of the Brazilian Amazon. Between 1960 and 1980, the population of Brazil's northern states more than doubled as settlers flooded into the region along an ambitious network of new roads. They found that once the trees were gone and the land exposed to baking sun and torrential rains, the soil soon degraded to the consistency of ground-up glass. Brush invaded the pastures, pests afflicted the agricultural plots, and the nutrients, which the forest had husbanded for millennia,

leached away. So did the settlers. In the Brazilian state of Rondonia, attrition is 82 per cent. Most get crops for only two or three years, pastures last not more than ten, and up to three acres are required to sustain one cow. This forces the deforestation front inexorably forward, as ranchers expand their holdings and settlers who have failed in one place go on to clear another. Behind the front lies an ever-growing expanse of abandoned and degraded lands. Susanna Hecht is a Brazilian specialist at the University of California in Los Angeles.

Susanna Hecht
According to the INPE, which is the Brazilian space institute, about 50 per cent of the areas that have been cleared are now abandoned or in some form of degradation. There are currently great battles raging over how much of the Amazon's been deforested, but let's take 15 to 20 million hectares as our sort of baseline data. You're looking at 10 million hectares in the last twenty years that have been reduced to rubble.

David Cayley
Cattle and conventional agriculture have proved to be completely inappropriate land uses in the Amazon region and, without the generous subsidies the government provided, uneconomic as well. The Brazilian government, under considerable international pressure, recently announced an end to its subsidies. But by itself, says Susanna Hecht, this will make little difference. There are now simply too many other forces driving deforestation.

Susanna Hecht
First, we can be certain that the Brazilians will continue putting in development there and that land values on the basis of continued governmental investment are likely to keep rising. Second, there's been an explosion in road development, so every time the Brazilians put in a road, they increase the value of nearby lands by several hundred per cent. Third, there are not many alternative investment areas in Brazil. The manufacturing economy is faltering. Farmers who have land elsewhere are often squeezed out as a function of consolidation of farms in southern Brazil, and land is still relatively cheap in the Amazon. Of course, you always sort of think that the problems

of soil were because other people were not quite as on the ball as you might be, so perhaps those could be overcome.

The other thing is that inflation right now is over a 1,000 per cent, so if you are a Brazilian with some money, where are you going to put it? You need to put it in something that can accompany inflation and that certainly is land, far better than almost any other investment.

So there are lots of reasons. The speculative gains to land are important, but the other thing is that just clearing it increases its value by more than 30 per cent, compared with forested land. So if you just clear your land, you can then turn it over at a far greater price. A lot of this land has been acquired illegally or without formal title, so essentially you get a free good and you can turn it into a marketed good. This represents a big one-time capital gain. Finally, there is timber that can be marketed. So there's been a lot of reasons that have made the Amazon very attractive, not for producing crops or livestock but for producing money, as a vehicle for capturing other kinds of resources.

David Cayley
Brazil is not the only country converting its jungles to pastures. From Mexico to Argentina, the cow has been the scourge of the forests. Ecologist Adrian Forsyth has watched the process in Costa Rica, where it was orchestrated by both the government and the international development banks. Ironically, the Costa Rican government has also made the strongest commitment to conservation of any tropical government, but its agricultural policies have still resulted in its being deforested virtually to the boundaries of its ambitious national park system.

Adrian Forsyth
The government decided that it wanted to build the beef industry, so it gave all kinds of incentives to create beef pastures, such as low-interest loans that you don't have to begin paying back for five years. This essentially means you have free money for five years, as long as you keep a beef herd. And so you can get a loan to buy more cattle, sell them off to the export market, get another loan, buy more cattle. The more cattle you

sell and the more loans you get, the more free money you've got, all without interest.

And so the government lending practices, which are supported by big organizations, say the Inter-American Development Bank or World Bank or whatever, just give entrepreneurs the incentive to go out and farm a huge amount of cattle. The cattle themselves aren't really worth much at all. You know, they're perhaps cheap beef for the importer, but the person who's actually doing the deforestation is making money out of this loan structure, where you essentially are getting free, no-interest money. That's always a strong incentive to get involved with anything.

If they'd given these kinds of loans for something else, say agricultural diversification, to get out of cattle and get into a small-area high-yield crop like asparagus or a tropical fruit or whatever, you would have had land use intensification, where instead of creating large, extensive areas of pasture, you focus on high-quality product in a small area. You would have had much less deforestation. Government incentives based on ignorance about land use have created a lot of the problems.

David Cayley
Costa Rica and the Brazilian Amazon are very different places. For one thing, Costa Rica's young, volcanic soils give the country far greater agricultural possibilities than the senile clays of the Amazon basin will ever allow. But in both places, the forests have gone at a terrible discount in relation to both their ecological and their long-term economic value. The tragedy of this is expressed in an affecting image from one of Englishman Adrian Cowell's films about the Brazilian Amazon. It shows a felled Brazil nut tree lying in a patch of scrubby corn. The standing tree might have produced 3,000 pounds of nuts, the corn only a meagre harvest for a couple of years. Surveying Costa Rica's Pacific slope from the Monteverdi cloud forest where he does research, Adrian Forsyth sees the same kind of waste.

Adrian Forsyth
In Monteverdi, for example, you look for miles and miles over these pastures, and they're brown and burnt-out for half the year by the dry season winds. And you see about ten cows in

this hundred square miles because the carrying capacity in the dry season is so low. And you look at the trees that they burned and there's at least thirty or forty tree species down there, several per hectare, that are worth far more than mahogany is, for example. It wasn't that they were logged and made into furniture for export to developed countries or anything like that. They were just chopped and burned. I've met *campesinos* who told me, "You know, I can remember cutting down *pocho'te* trees, the most valuable lumber trees in all of Costa Rica, just to get a fistful of honey out of a bee's nest." It was a complete and utter waste of the resource because no one could ever conceive that they would run out of trees.

And so, in this area that's supporting a hundred cows, you see literally the skeletons and burnt-out remains of billions or at least millions of dollars of board feet of lumber. If these guys had just cut this stuff and warehoused it, they'd all be rich men today. But that's the whole thing. It's not making people rich, it's making them poor. The true tragedy is this complete throwing away of a valuable resource.

David Cayley
Throughout history, rain forests have advanced and retreated with the rise and fall of civilizations. Today, civilization scours the entire planet, and the rain forests only retreat. We are witnessing the last assault, and, according to tropical ecologist Dan Janzen of the University of Pennsylvania, it's different in kind from anything that has ever happened before.

Dan Janzen
Many big tropical areas have been under heavy human agricultural impact for millennia. Today, we're losing the species from those areas. For thousands of years, we did not. What's the difference? The difference is that, for thousands of years, we cleared and damaged and trashed in a mosaic pattern, and the species kept moving back and forth between the spots that were relatively less damaged. That's what shifting agriculture is. Today, we come in and we clearcut. There is no historical analogue to a 100,000 square kilometres of cattle pasture in Brazil. All the Yucatan peninsula was agricultural land 1,000, 1,500, 2,000 years ago — all of it. Today, you fly over it and it's unbroken wilderness, except if you use satellite photographs,

you can see the dikes and canals underneath the trees. We have big areas all over the tropics that at one time sustained enormous populations and were very thoroughly cleared, but have now regenerated back to forest. The difference is, we let them regenerate back to forest.

David Cayley
Today, with no prospect of regeneration, the world faces the possibility that there will soon be no rain forests outside of protected parks and reserves. This is bound to have ramifications that go far beyond the tropics. It will certainly result in mass extinctions, and it may also have implications for the world's climate.

Jose Lutzenberger
Tropical rain forests are not, as some people like to say, the lungs of the planet. That analogy is wrong because lungs don't produce oxygen, they consume oxygen. The rain forests are a fantastic and incredible climate machine. They are, so to say, the air conditioner of the planet. Straddling the equator with their fantastic evapo-transpiration, they are a heat sink. They cool the atmosphere of the planet, and they affect both hemispheres. The rain forest has an evapo-transpiration of about 75 per cent, meaning that the rain that falls on the forest goes back into the atmosphere. Some 20 per cent of the rain water that hits the trees never hits the ground. It wets the leaves and is re-evaporated back into the atmosphere. Of the 75 per cent that reaches the ground, only about 25 per cent ends up in the streams and rivers and goes to the ocean. Even on the way to the ocean, part of it is re-evaporated in the atmosphere and the rest is absorbed by the plants and put into the atmosphere by transpiration. That's why we say "evapo-transpiration" — evaporation and transpiration. The rain forest makes its own climate.

David Cayley
Jose Lutzenberger's theory has a frightening corollary. Reducing the forested area beyond a certain critical but unknown point could lead to irreversible drying. This would heat the earth at the equator with unpredictable consequences for the planet as a whole.

Adrian Forsyth
When you eliminate that vegetation, you create this arid zone. You can see it over large Amazonian cities. There will be big cumulus clouds stacked up over the forests and as soon as you get to the city, the sky is absolutely clear. The ground is very hot because there's no vegetative covering and it gets baked. And so, if you eliminate vegetation over an area almost the size of the continental United States, you essentially turn off this big equatorial evaporative system, which, undoubtedly, is going to change all the rainfall patterns north and south of that system.

David Cayley
In addition to its uncertain influence on the climate, tropical deforestation plays a role in the greenhouse effect — the build-up of so-called greenhouse gases like carbon dioxide in the atmosphere, leading to a gradual rise in global mean temperature. Forests are what scientists call a carbon sink. They take carbon out of the atmosphere when they grow. When they are burned, as forests cleared for agriculture usually are, this carbon is released. Burning of tropical forest now contributes between 20 and 40 per cent of the carbon pumped into the atmosphere annually. It could, in other words, make a critical difference in a problem that tropical ecologist Tom Lovejoy fears could become a runaway. Lovejoy is Assistant Secretary for External Affairs at the Smithsonian Institution in Washington.

Tom Lovejoy
I think most scientists would agree that the earth on average is already a half degree centigrade warmer. It doesn't sound like very much, but then remember the difference between the glacial period, when most of Canada was under ice, and an interglacial was only five degrees centigrade. So what we're talking about is already a change equal to one-tenth of what made such a massive change in the planetary ecology, except it's in the other direction. On top of that, the earth doesn't instantly warm up when more carbon dioxide is put up in the atmosphere. It's an accumulation. So we're already committed to some more warming. Then on top of that, you have to realize that the warming will be uneven, that the farther you

go towards the poles, the greater the warming will be, which may seem great in the Yukon in the middle of January, but what it will do is massively alter the ecology. The snow line, as it were, will change. As soon as you no longer have snow up there, you will get further warming because the white snow won't be there reflecting the energy back, and there's a real possibility of a runaway effect.

Adrian Forsyth
The thing that's emerging about the greenhouse effect is that it's a positive feedback system, so that a small addition of carbon dioxide may encourage further additions. So that 20 per cent, say, contributed by tropical deforestation may only sound like one-fifth of the problem, compared to fossil fuels, but that 20 per cent may really be crucial if it kicks off another change. For example, that 20 per cent rise in temperature could cause all the peat moss in the boreal forests of Canada to oxidize, because it was just at that stable point, and the 20 per cent kicks you over that stable point. You release all the carbon from the northern bogs and tundra, and you're into another problem where it's adding methane to the atmosphere.

And so, it goes on and on, and scientists believe that's how the ice ages work. They believe that a positive feedback cycle occurs where things go way past equilibrium and take a long, long perturbation before they start to reverse, and why they reverse is anyone's guess. So I think tropical forests are poised to play a major role in that process, either negatively by being burned and being turned into atmospheric carbon, or by being reforested and turned into an incredible sink, which will slow down the process.

David Cayley
Fossil fuel burning is still the major cause of global warming, but burning of tropical forest may be a fairly close second. According to Tom Lovejoy's figures, approximately five billion tons of carbon are put into the atmosphere annually. Two billion are removed by natural process, absorbed by the oceans or growing plants. That leaves three billion. Last year, in Brazil alone, burning of forests put 600 million tons of carbon into the atmosphere, which is 20 per cent of three billion. If the global rate was double that, which seems a reasonable as-

sumption, then deforestation may be contributing as much as 40 per cent of net atmospheric carbon increase. It follows, for Dr. Lovejoy, that a combination of reforestation and ending deforestation could make quite a difference.

Tom Lovejoy
If one was able to reforest in various places around the world, north and south — I mean, I think every country has to participate, at least every country that can grow trees — and you could reforest on the order of one to two million square kilometres, you could remove about a billion tons of carbon from the air. So there's a potential there, playing with reforestation and controlling deforestation, to bring the imbalance in the equation down from three billion tons per year to maybe one and a half billion tons a year, and that can leave energy conservation, energy efficiency with a reasonable number to work on to try and bring it down to zero. But — and here's the big *but* — this would only work for approximately thirty years because when forests begin to mature, they cease to be a major uptaker or sink, in the scientists' term, for carbon. So you have to determine what to do after that. I think the way to look at it is that by playing with energy conservation, efficiency and management of forests we can buy thirty years to develop a new energy scenario for society.

David Cayley
This implies management on an unprecedented scale.

Tom Lovejoy
Without any question. We have to start managing the planet as a whole, which is not an easy leap for a bunch of sovereign nations. And this is not going to be solved on a business-as-usual basis. That's what it comes down to. I think we have less than ten years to get it together. I think we need to approach this with the same sense of urgency and the same psychology as if we were on a war footing, and the only difference here is that we're at war with our way of life and with ourselves.

*** *** ***

Tropical deforestation has extinguished species, jeopardized climates, and possibly increased the instability of the earth's atmosphere. In many places this irreplaceable heritage has simply been thrown away without any offsetting benefit. Through deforestation, societies have grown poorer, rather than richer. And, as the next chapter relates, this has had unconscionable human, as well as ecological, costs.

5

The People of the Forest

Around the world there are an estimated 500 million people who still depend on forests for at least part of their subsistence. Most of them are now threatened by the rapid pace of deforestation. The second programme of "From Commons to Catastrophe" dealt with resistance to rainforest logging: the preservation movement in Australia, the desperate last stand of the Penan people of Sarawak, and a successful campaign to expel a logging company from a community in the Solomon Islands.

David Cayley
Paradise is a village on the Pacific island of New Georgia, one of six large islands which make up the Solomon Islands, a former British colony that lies northeast of Australia. Here in the early 1980s, a battle was fought between the villagers and Lever's Pacific Timbers, a subsidiary of the giant British-based multinational Unilever. In 1978, Lever's applied to log 75,000 hectares of pristine rain forest on north New Georgia. Lever's had been logging in the Solomons for twenty years and virtually written the rules for how it was done: clear felling of all saleable trees, with no replanting. Eighty per cent of the land in the Solomons is communally owned and the affected communities were split on whether Lever's should be allowed to

proceed. Job Dudley Tausinga was then a member of Parliament and one of those who opposed the logging, because he realized that the forests were irreplaceable.

Job Dudley Tausinga
The people and the forest are interdependent. The Solomon Islands for a long time to come will still use the forest. The medicines, their food, the building materials. It is a source of life for the people in the Solomon Islands, and whether we like it or not, even if we have to try to come up with modernization, the people will still be dependent on the forest.

David Cayley
With the community split on whether logging should be allowed, the government intervened on behalf of the company. It passed a law that affirmed the community's ownership of the land, but vested ownership of the trees in a state corporation, which then authorized Lever's logging plan. This legal trick intensified local opposition. When all negotiations failed, the men of Paradise village decided to settle the matter in what they called "custom style."

Job Dudley Tausinga
When the government and Lever's did not listen to the people, the people then made some peaceful demonstrations or peaceful protests, but that again did not work and the people were put in jail. There is a limit to everything and the patience of these people was running out, so consequently they raided the company's camp at Enoghae.

David Cayley
Just before daybreak, on March 26, 1982, 150 men set sail from Paradise for the Lever's logging camp at Enoghae. They landed out of sight of the camp and approached through the forest, by night. Their bodies were smeared with the traditional war paint of mud and leaves. Green vines were tied around their heads, arms and legs, and they carried palm frond torches to light their way. At dawn, they burst into the logging camp. The workers and their families were ordered out of their quarters so that no one was hurt, and then seventy-eight houses, the company store, three bulldozers and the

crane were smashed and burned. The damage was over one million dollars.

The raid on the camp at Enoghae was the beginning of the end for Lever's in the Solomons. Of the fifty men arrested in connection with the raid, only seven were ever jailed. When the company tried to land bulldozers at Enoghae ten months later, the wharf was set alight and burned. Meanwhile, resistance was spreading. In 1984, another logging camp was burned, this time by the people of a different village. In February, 1986, another of the company's bulldozers was damaged, and finally, in October of that year, Lever's sold its assets and left the Solomon Islands altogether. Behind it, the company left tens of thousands of hectares of land reduced to nothing more than a vine cover, lands that will take hundreds of years to recover and thousands to be anything like what they once were. After eight years of struggle, peace finally returned to Paradise. Faced with the company's appetite for logs and the government's appetite for royalties, a determined people had put their own subsistence first, and won.

Job Dudley Tausinga led the campaign throughout, and in the midst of it, in 1983, he was elected premier of the Solomon Islands western province, which included New Georgia. He looks back now with satisfaction.

Job Dudley Tausinga
After all the work we had done, after all these negotiations, these confrontations with Unilever, we were finally free to have the normal life that we used to have. I was born in the forest and I still love it, and I think the problem I have, or the people have, is to appreciate the usefulness of the forest. We need to create some kind of awareness so that people can see that the forest cannot exist without the people and the people cannot exist without the forest. If there are no people, then no one will say, "This is a forest."

David Cayley
The battle between Lever's and the people of New Georgia is only one of many similar struggles going on around the world. The action may have been dramatic and the outcome unusual, but the essential conflict between those who depend on the standing forest and those who depend on cutting it down has

been repeated again and again. It has been particularly acute in Southeast Asia, where a careless cut-and-run mentality has run through country after country like a wildfire. The Philippines has virtually exhausted its productive lowland forests. At the current rate of exploitation, Malaysia has maybe ten years of log exports left to go. By the time Thailand declared a logging ban in December of 1988, after two years of disastrous deforestation-induced flooding, it had already become a net importer of wood, and Indonesia now deforests an estimated 700,000 hectares annually.

Why these forests have been so badly managed is a question that interests Theodore Panayotou, a Canadian citizen who is a research associate of the Harvard Institute for International Development. He says the main reason is that there has been no one in a position to manage them properly.

Theodore Panayotou
Until the Second World War, most forests in the world were owned by individuals or by communities. After the Second World War, governments in tropical countries, imitating Western models, appropriated forest resources and declared state ownership without accommodating the rights of local communities who had had access to these forests before. Moreover, the governments did not have sufficient enforcement capability and sufficient presence in those forests, and basically what occurred was a conversion from well-managed, traditional, communal forests into unmanaged, open-access, everybody's property, state forests.

David Cayley
So the forests were left without effective guardians, such as the people of New Georgia, who retained communal rights, proved to be for their forests. The governments who appropriated the forests were generally interested in development at all costs, an attitude exemplified by an ad which the government of the Philippines placed in *Fortune* magazine in October, 1975. "To attract companies like yours," President Ferdinand Marcos asserted in this ad, "we have felled mountains, razed jungles, filled swamps, moved rivers, relocated towns, and in their place built power plants, dams and roads, all to make it easier for you and your business to do business here." Thus,

the forests were parcelled out to national and international logging firms, typically in short-term concessions. These were intended to preserve the long-term interests of the state, but they had a perverse result.

Theodore Panayotou
Concessions are less than twenty years and, as a result, since it takes more than forty or even seventy years for trees to regenerate in the tropics, concessioners have no interest in the second or third crop. Instead, they go in their concession and they cut and run, because their concession will expire in a few years and if they conserve anything, they have no assurance that their concession will be renewed in order to take advantage of whatever investments they make in preserving the productivity of the forest, in replanting, in regeneration, et cetera.

David Cayley
The problem was aggravated by the fact that the governments concerned usually charged less for the resource than they could have. Just how much less was made clear by a recent World Resources Institute study called *The Forest and the Trees*. The authors of this study worked out how much governments had taken of what was left over when all costs of logging, including a reasonable profit, had been allowed, what economists called "rent," or the part of the price that properly belongs to the owner of the resource. They found that between 1979 and 1982, for example, the government of the Philippines took only 16.5 per cent of these rents and the government of Indonesia only 38 per cent. The rest was left in private hands. According to Theodore Panayotou, this had two results. It encouraged loggers to cut in marginal areas, which would not otherwise have paid, and it accelerated the rate of exploitation.

Theodore Panayotou
Concessioners feel that this is a very unusual situation. They don't usually get that amount of free money, and when they get it, they realize that this is going to be temporary, that the governments one day will realize what's going on and they will change the terms of the concessions. Therefore, what would you do? You would try to make hay while the sun

shines, and try to cut as many trees as possible while you are still getting a big share of rents.

David Cayley
The foregoing of rents, along with the short terms of concessions, have created what amounts to a fire sale in the forests of Southeast Asia. According to the World Resources Institute study, the Philippines government's logging revenue share did not even cover its related infrastructure and administration costs. The result is that country after country has given away its forests at less than the cost of replacement.

Forests provide other amenities than timber. They hold soil, regulate the flow of water, control local microclimates and provide fruit, game, medicines and other non-timber products to those who live in and around them. Some of these so-called "minor" forest products also enter into trade. By the early 1980s, Indonesia had export earnings of some $120 million a year from the products of the standing forest — resins, essential oils, medicines, rattan, flowers, and so on. The potential of such products is unknown, since the economic value of the standing forest has generally been overlooked. Conservation now depends on a new economic calculus that considers the value of the forest as a whole.

Theodore Panayotou has just co-directed a study for the International Tropical Timber Organization which proposes that tropical forests be managed with a view to all their uses — "multiple use management" the study calls it. He says that a first step towards a more holistic management would be to return control of tropical forests to the communities who live in them.

Theodore Panayotou
Under tropical conditions, in developing countries, if there are people who live around the forest or in the forest, they are probably the best suited to manage those forests because they have a specialized knowledge of their ecological properties. Moreover, by their physical presence and by the direct connection between their livelihood and the long-term productivity of those forests, they would be most likely to manage those forests for the maximum long-term value.

David Cayley
Along the northwest coast of the island of Borneo lies the state of Sarawak. Once it was ruled by a British family, who styled themselves the white rajahs of Borneo. Then it became a British crown colony, and today it is one of the eastern provinces of the Federation of Malaysia. It is home to some half million native people, and in the last twenty years, they have suffered cruelly from the depredations of logging companies in their lands.

One of these peoples is the Penan, a small group who have traditionally lived by hunting and gathering in the forest. Because they are nomads and because the state government of Sarawak aggressively discourages foreign journalists, they are not always easy to find. But in 1988, two Australian film-makers, Jeni Kendal and Paul Tate, managed to meet with one of the last groups of nomadic Penan.

The interviews with the Penan that follow are taken from their film, *Blowpipes and Bulldozers*. The translator is Bruno Manser, a young Swiss who came to Sarawak in 1984, and went to live with the Penan, adopting their ways, making hundreds of drawings and compiling a first-ever Penan dictionary. He left Sarawak in the spring of 1991, because he felt that his own notoriety was detracting from the Penan cause.

Bruno Manser (translating for Penan man)
He says that the company entered these grounds maybe four, five or six years ago, and this makes our life hard, for we take water from the river, we take our fish from the river. All of our food we find in the jungle, and this is all being destroyed. The water is dirty, the wild boars and the wild game are fleeing or are being killed by the loggers, and we are now like fish thrown on the land. The jungle, all the trees, they are like our home, like our house. Even the little trees, we need them. A few of us follow the company, and are corrupted. The company knows how to make the dollar sweet to the mouth, but most of us, we don't eat dollars, we eat the heart of the palm, we eat the wild boar, we eat sago. When we see our lands getting destroyed and we hear the noise of the bulldozers destroying our land, how can we not get sad and angry?

David Cayley
In the last couple of years, the situation of the Penan and other Sarawak natives has begun to attract increasing international attention. A recent visitor who went to see for himself was Canadian conservationist Mat Sylvan. The name is a pseudonym used at his request in order not to compromise his continuing work on behalf of the Penan. He returned to Canada literally gripped with horror at what he saw.

Mat Sylvan
I don't think it's too extreme to say that there's a cultural and biological holocaust happening in northern Borneo. The forests of Borneo, like the forests of West Malaysia and Sumatra, are the oldest living ecosystem on the planet. They're 130 million years old and are the richest and most diverse ecosystem we have left on the planet. When much of equatorial Africa and South America were becoming dry savannah lands during the great ice ages, that part of Southeast Asia kept its microclime, kept its warm climate and its moisture levels, and continued on a sort of uninterrupted evolution.

So this is a world resource, a world treasure that's being liquidated at the most rapid rate of any rain forest area on the planet. Between 60 and 70 per cent of the forests of Sarawak are gone. They've been logged. There's very little left and progressively, over the last 25 years, they've been moving from the coast up in towards the mountains, close to the Kalimantan border, and just systematically destroying it.

And it's really speeded up in recent years. When you go up the huge rivers draining out of the highlands of central Borneo, you feel like you are in the closing scene of *Apocalypse Now*. All you can think of is the horror. On both sides of the river, lined up to 40 feet high, are stockpiled logs that have just been cut and are sitting there. The rivers are jammed with barges going out with logs and logs and logs. You get to the mouth of the rivers and Japanese freighters are lined up as far as you can see on the horizon, taking the logs out. The scale this is happening on and the speed with which it's happening is colossal.

David Cayley
The communities involved regard what is going on as theft. Sometimes the logging company surveyors just literally "turn

up," and that is when the community learns that the government has given their land to a logging company. They have expressed disbelief at the way they're being treated. Bruno Manser has heard the complaints of the Penan, with whom he lives.

Bruno Manser
Of course, they look at it as an injustice because they know that their fathers and grandfathers have been living here for years. They reply, for instance, when they are angry, Why don't you listen to us? If you come to our woods, then you should also allow us to go down river to your shops and just help ourselves, and then you shouldn't punish us, because that's what you are doing with our lands.

Mat Sylvan
These people have had everything they've ever known stolen from them. They've had the land ripped out from under them. The only place they know to get food is the forest. There's no longer a forest. The only home they've ever known has been the cool shade under the forest. There's no longer that. They're out in the scorching sun. The only thing they've ever known in terms of a society is a few families together. Now they're forced together, three or four hundred people. The diseases are spreading. The water is so polluted from the logging, the runoff of the topsoil, that people are getting very, very sick drinking it. There are no fish left in the rivers. When these people get sick, they might have to walk for five days to find a stand of forest where they can find a plant to cure themselves. And as I mentioned, many of these diseases affecting them are things they've never known. They don't know how to deal with them. A Penan man and I were looking at an area that had been cut 25 years earlier, and I said, Surely there must be some plants coming back here that you could use as a medicine? And he said, Well, I'll show you one, and he took me to a tree that was about three feet high, and he said, This tree now has six leaves on it. It has enough leaves that we could take all these leaves and use them for fever or for diarrhea, or whatever disease it was for. He said, But if somebody was sick, how could we wait 25 years for our medicine to be here?

David Cayley
The logging practices, which have created this damage, are an illustration of the difference between theory and practice in tropical logging. Theoretically, Malaysia practices what it calls "selective cutting." What actually happens on the ground, says Mat Sylvan, is closer to total destruction.

Mat Sylvan
The government of Malaysia says their forestry practices are excellent. They only take 40 per cent of the trees out of the woods. The rest are left there, so the forest is always still there and intact. Well, it's not true at all. Much closer to 70 or 80 per cent of all the trees are dropped and those that are left are usually seriously damaged. The whole microclimate is gone. There is no longer a moist, dark area under the forest canopy for wildlife. It's just open winds coming through and the area becoming drier. There's incredible flooding going on there now. I had to wade out of the last village I left. I was up to my chest in water because 80 per cent of all the rainfall that comes down there, and it rains a lot in Borneo, through transpiration and evaporation goes up through the trees and goes back and forms cloud cover again. Well, with 60 to 70 per cent of the area deforested now, most of that rain is staying on the ground, and the very, very rapid runoff on the surface is taking most of the topsoil with it into all the tributaries and all the major rivers. So the rivers are filled with sediment. I've never seen a higher sediment load in my life than what's going down those rivers in Sarawak.

David Cayley
Logging damage in Sarawak has been so severe and the pace of destruction so swift, that many of the Penan have already been forced to settle in permanent villages. They have been encouraged by a government that sees their way of life as an embarrassing anachronism, but they have proved to be only half-hearted agriculturalists and still depend on the forest. These communities are now deeply frustrated and dispirited.

Penan man (speaking through translator)
Before, when we first came in touch with the government officials, we were advised to go and settle near the riverbanks

so we could live in a longhouse, in a community, and it would be easier for us to be in contact with others and they could slowly teach us to lead a better life than what we were used to in the jungle. But now we are here, this is what we see. The company is coming here now and destroying everything we have. They treat us very badly, and if they continue to treat us like this and never listen to what we say, something bad is going to happen and we are not to be blamed. We have tried to discuss things with them, but they never listen to us.

David Cayley
The Penan who remain in the forest maintain their traditional way of life, hunting small game with their blowpipes and harvesting the sago palm. But they have been steadily retreating before the advancing loggers and now, according to Bruno Manser, their backs are against the wall.

Bruno Manser
In five or six years, this area will be logged, from all sides. It won't take long. It may take ten years, at the most.

Interviewer
Then what happens to the Penan?

Bruno Manser
Of course, they get civilized.

Interviewer
And they move to where?

Bruno Manser
They have no possibility to move anywhere again. That's the way of the nomadic tribes in the past. When the company comes, then they all go to another place. But now the companies are like scissors. They are locked between and they can't go to where there is still primary jungle.

Bruno Manser (translating for Penan man)
Within one year, or in the year after that, if the company cannot be stopped, it will be too late. Maybe only three months, if the company really drives on the logging roads. If

you walk on foot, you can cover our whole land within one week. It's not a big land. But if our land is destroyed, that's like our death.

David Cayley
The destruction of the lands of the Penan is happening in spite of a clear recognition in Sarawak law of their communal rights. According to the state constitution, native people own the lands that they have traditionally occupied and used. Unfortunately, says Martin Khor of the Malaysian branch of Friends of the Earth, what these lands are has never been defined.

Martin Khor
There was no document given to each community to demarcate what is considered their customary land, so it is a grey area. These timber companies get a document or a licence from the government, which includes a map, and indicates their concession area. So here you have a dispute between the timber company who possess a document and a map, and native peoples who say that these are our customary lands that you have intruded upon. In a conflict like that, the one who possesses a document and map tends to get the upper hand, and the natives find themselves at the losing end. Though they possess rights, legal rights to their customary land, the definition of what that is is often a grey area and they suffer because of it.

David Cayley
A prime reason why these rights have not been respected is because a lot of people are getting rich by abusing them. In Sarawak, logging concessions are one of the perks of political office. The environment minister, James Wong, owns one of the country's most prosperous timber companies, Limbang Trading, which has over 250,000 acres of timber concessions. During the recent Sarawak election, it was revealed that the chief minister and his predecessor between them have a hand in an astounding 30 per cent of the country's forest land. Community after community has tried and failed to impress their demands for communal forest reserves on the state and federal governments. By the early 1980s, there were already sporadic outbursts of violence. In 1982, a militant Iban community

drove a logging community from their lands by destroying bulldozers and logging trucks, and by 1987, frustration was so general that all the affected peoples decided on concerted action.

Martin Khor
It started many years ago, on an individual basis, when people started putting up little barricades of their own, here and there, in one or two communities, to try to prevent the loggers from coming in. And then, of course, they have also been writing petitions, have been seeing government officials, and so on, but to little effect. So it was sometime in March and April, 1987, that twenty different communities in one concerted effort, put up blockades in their own areas and managed to freeze logging activities. Representatives of the native leaders also went to the state capital of Kuala Lumpur to meet various government officials and ministers, to appeal to them to put an end to their plight.

David Cayley
The appeal proved fruitless. Police descended in force on the blockades and broke them. Forty-three Penan men were arrested, jailed and charged, adding to the suffering which they had already experienced in maintaining the blockades.

Mat Sylvan
I was told that people in some of the places had starved at the blockade sites because they were so far away from any food sources. We were told of a woman who was actually run over and crushed by a bulldozer, and yet the people have not struck back. Despite provocation on the part of the military and the police, they have not struck back. They have been arrested and peacefully been taken to jail. We also heard of some extreme human rights violations going on there. The men have been handcuffed together in a fashion that they actually have to urinate on each other. They've been put into the toilet sections of the prisons with other criminals. They've been thrown small, rotten fish as the only source of food. We were hearing these stories from many different sources, many different villages, isolated over long distances, and you can't help but feel

that there's some truth to them, as much as the government of Sarawak tries to deny them.

Bruno Manser (translating for Penan man)
We made a blockade. But we lost, for the police came and destroyed our blockade and burned it down. We just stood by. We didn't do anything. We could just look at how they destroyed our blockade. And now we get treated not like human beings, but just like short-tail and long-tail macaques. They don't listen to us. We are afraid now. We are afraid when we try to stop the company, then we will be arrested, but we don't have any *ringit*, we don't have any dollars. Who will bring us out from jail? Who will feed our women, our children when we are in jail?

David Cayley
Despite the privations involved, the blockades have continued. Sarawak is now under virtual martial law. A leading Kayan activist, Harrison Ngau, was jailed for sixty days without trial and is now under house arrest. Journalists have been kept out of the country and away from the blockades. In January of 1989, 127 more Penan were arrested. They faced two-year jail sentences and $2,500 fines under a new section added to Malaysia's forestry ordinance late last year. Resistance continues among the other native communities of Sarawak as well. What is at stake for the Penan is everything. What is at stake for those of us who live inside technological civilization is the continued existence of possibilities other than our own.

Bruno Manser
What's most impressive is the social context in which the Penan live, that they share everything they get in the jungle. I am embarrassed when I think about our civilized people, how we share what we get. We are really big egotists, and the Penan, I have enormous respect for them.

Mat Sylvan
One of the experiences I noticed with the Penan people, is this instant sharing, this unconscious act of sharing that takes place. If you hand some food to some people or even see them

come out of their forest with some food, it's automatically passed to every single person in the group, without any thought of "me." They never think of "me first." We saw this sharing in the Penan villages all the time, and the people had very little, maybe just half a cup of sago palm for each person to eat a day, yet we would get our portion. And I was very moved, one of the last days I was there, when a woman asked me, in the presence of a translator who could translate Penan into English, "Where could we go in your country and be fed and welcomed in the home of a stranger?" She thought that they would be treated with the same generosity in our country and I was really taken aback by that. I thought about it and decided you probably couldn't.

David Cayley
The principal beneficiaries of logging in Sarawak are the local logging companies who do the cutting and the big Japanese timber concerns that import most of the wood in the form of raw logs. For nearly twenty years, Japan has been the world's largest importer of tropical timbers, husbanding its own forests while benefitting from a steady supply of cheap, high-quality hardwoods from the old-growth forests of the Philippines, Indonesia, and now Malaysia. In some cases, it has also financed the logging. In 1987, 96 per cent of its imports came from Sarawak, the neighbouring Malaysian state of Sabah, and Papua, New Guinea.

Randy Hayes of the Rainforest Action Network in San Francisco recently visited Sarawak with a U.S. delegation that included congressional staffers and environmental and human rights groups. He thinks that international pressure to end overcutting and the trampling of native rights in Sarawak, which has so far failed to sway the Malaysian government, should now focus on Japan.

Randy Hayes
Japan and the Japanese yen have become kind of the Darth Vader of the tropical rain forests and the tribal peoples of the world. Consequently, we have switched our strategy to help curtail the timber cutting in the tribal regions by pressuring Japan, asking them to suspend the purchase of timber from the Sarawak regions that are contested by the tribal groups. We've

been pressuring them by launching an international letter-writing campaign and to date, we have about 57,000 signatures. Fifty per cent of those signatures are Japanese citizens, so there's an uprising in Japan of concern, to the point that 20-some thousand people have been willing to sign a petition that was presented to the prime minister's office on April 24, 1989.

David Cayley
This is the first time that Japan has really experienced heat about its role in tropical deforestation. Japanese tropical log imports are also the subject of a recently published World Wildlife Fund report called *Timber from the South Seas*. This report got front-page treatment in the Japanese press and even resulted in questions to the prime minister in parliament. Concern focuses not just on the volume of imports, but also the uses to which the wood is put. Peggy Hallward is director of forestry research for Toronto's leg of the world rain forest movement, Probe International.

Peggy Hallward
The saddest thing about the wood from Asia is that it is used as cheap construction material for mouldings for concrete for the housing market in Japan. If this wood was at least used for veneers, then people could enjoy its beauty and pay a fair price for it. But we're not even talking about that, we're talking about plywood that goes into floorboards and plywood that is used to make mouldings for concrete. The mouldings are used three or four times and then they're discarded. So we're talking about an incredibly sad way to use the most valuable woods in the world, when alternatives exist. We don't have to use these particular woods. There is so much degraded land around the world where we could be replanting trees to use for construction. We don't have to be cutting down primary rain forest.

David Cayley
The tropical timber trade has now become a major focus of environmental action. In Germany, thirty local councils have banned the use of rain forest timbers in building projects and two hundred more councils are due to vote on the issue this year. In Australia, protestors have blockaded ships carrying

Malaysian logs into Sydney. Canadian conservationist Mat Sylvan thinks that these kinds of actions are essential but should be kept in context.

Mat Sylvan
We have to launch a global campaign to put pressure on Malaysia to develop protective policies, but we have to do it with a good deal of humility. When I spoke to Malaysian officials, the one thing they can't stand is people from countries like ours, that have historically deforested our continent, that have treated our native people in much the same way the Penan are being treated, being righteous about this. We're in no position to be righteous. The deforestation that's going on in Canada is the most appalling temperate rain forest logging in the world. And Malaysia can say, Look, we've set aside 6 to 8 per cent of our entire nation as national park. Indonesia can point to 9 per cent and they're working for 18 per cent by the year 2000. Canada has got 1.7 per cent of its land mass as national park. Here's one of the wealthiest countries in the world, second largest land area, lowest population density, no crushing foreign debts like Malaysia or Brazil, and we take thirteen years to save one-tenth of 1 per cent of our forest base, like South Moresby — one-tenth of 1 per cent of the B.C. forest base. It's infinitesimally small on the national scale.

What we must do is say, Look, we realize we've got problems, we've all got problems. Canada's got to start setting a model for the world for environmental protection and forest protection, and say to the world, How can we help you with the situation?

David Cayley
Most of the battles to save rain forests have been fought in countries like Sarawak and the Solomon Islands, where people still depend on the forest for subsistence. But at the end of the 1970s, protests also erupted in an industrialized country, Australia, where a powerful conservation movement fought a series of pitched battles with loggers and police and eventually managed to protect much of the country's remaining rain forest. Once, most of eastern Australia was covered by tropical and subtropical forests. By 1979, when John Seed got involved,

only a few remnants remained, and he and his friends determined to save them.

John Seed
I'd no interest in ecology whatsoever until 1979. I mean, I read the newspapers, I was concerned, but I wasn't a member of any organizations. It was totally alien to me. I was a meditator, basically, and a grower of tropical fruits and subtropical fruits on what I later realized were the margins of what had been the largest subtropical rain forest in the world, known as the Big Scrub. There was just a tiny fragment, less than 1 per cent of this, remaining. Neighbours of mine had been struggling, unknown to me, for four or five years to protect this last remnant, and it just came to my attention in 1979 at one of the local produce markets that were held regularly in that area. Someone got onto the microphone and appealed for help to protect this forest. And I still can't remember what actually persuaded me to take part, but a profound change took place as a result of experiencing that forest, especially in the conditions of danger, in the excitement of a civil disobedience action where people were lying down in front of bulldozers and getting arrested by the score.

David Cayley
The action happened in an area of New South Wales called Terania Creek over a period of two years. The first blockades resulted in a logging freeze and a government inquiry. When the inquiry proved unsatisfactory, the protests resumed in the fall of 1982. Hundreds of conservationists were arrested in the skirmishing which followed, but the government finally acceded to their wishes and created a series of parks that protected more than half of the remaining rain forest in New South Wales. The premier of New South Wales, Neville Wran, called the battle for Terania Creek a turning point.

Neville Wran
It took Terania Creek to focus our attention on the fact that first of all, trees can't be replaced, that they are such an important part of our natural heritage, but also that there were people, not in the hundreds, but in the hundreds of thousands, who loved those rain forests and were prepared to get out and

fight for them. Politics in terms of the environment has never quite been the same.

David Cayley
Strengthened somewhat by their victory at Terania Creek, Australia's growing conservation movement next shifted its attention to the Franklin River in Tasmania, where a hydro dam was about to flood the heart of Tasmania's temperate rain forest wilderness.

John Seed
The Franklin River became the largest environmental confrontation in Australian history before or since. About 3,000 people came from all over to this very remote corner of the country to take part in this blockade, and more than 1,500 were arrested. The blockade would typically be a string of little, puny, yellow rubber inflatable craft stretched across the river with the whole might of the Tasmanian police force in big barges being towed by tugs with earth-moving machinery on the back of them, breaking through these blockades.

But there were also helicopters and boats full of reporters, so these images were rapidly communicated around Australia, and we'd timed the blockade to take place a few weeks before the federal elections in 1983. The interest all around the country, rallies held in every capital city around the country, with thousands of people in support of our cause, prompted Bob Hawke, the leader of the Labour party, which was then in opposition, to promise that if elected, he would stop the dam. And this is what we'd been waiting for. At this point, we just left a skeleton crew down on the river and fanned out to eleven marginal electorates that we'd identified around the country where less than two per cent separated the two political parties. We lobbied intensively in those electorates, knocking on every door twice in the week preceding the election. All eleven of those electorates swung to the Labour party, and Bob Hawke's first words upon being elected were "The dam shall not be built."

David Cayley
The next threat to the rain forest came in northern Australia, where the North Queensland government decided to drive a

road through one of the last remnants of Australia's most ancient forest, a forest in continuous, undisturbed evolution for over 100 million years. North Queensland premier Martin Tenni bid defiance to anyone who dared to try and block his government's plans.

Martin Tenni
We will bring bulldozers in there, if need be, to clear this road, and it will be needed. We will cut the tops of the ranges down if it's needed, and no hippie, no greenie, no environmentalist will stop that from happening. They can try their hardest, they can do what they like. They won't win.

David Cayley
Australia's "greenies," as Martin Tenni called them, rose to the challenge. They were determined to block the road through the rain forest to Cape Tribulation, and to do so, they devised even more dramatic tactics than they had used in their earlier struggles.

John Seed
What we'd done was dug holes at the beginning of their proposed road at a point where they couldn't outflank us. We cast cement slabs in the bottom of these holes with steel reinforcing and with high tensile steel chain coming out of the reinforcing and the slabs. When the bulldozers came, we jumped into the holes, chained our ankles to the cement in the earth and then filled the holes up, so that only our heads were sticking out of the ground when they came.

This slowed them down for two or three days. During these two or three days, there were images of people buried in the earth to protect the rain forest. People were in a very vulnerable situation, especially some people who left their hands buried as well and their arms, so that they couldn't raise their hands to defend their face. That was a very moving kind of image, and this burned itself onto the retina of the Australian television viewing public.

The police eventually turned dogs loose and people were hospitalized. It was very, very rough, and they managed to push us aside, of course, and they completed the road. But Daintry, Cape Tribulation, this area became a household word in Australia and now, a couple of years later, the wheels have slowly turned and it looks like this area too is going to become world heritage. In spite of the objections of the Queensland government, the federal government looks like it's pushing this through.

David Cayley
Through the courage and imagination of Australia's conservationists, much of the country's remaining rain forest is now protected. Activists like John Seed have subsequently turned their attention to the world, supporting the struggles of people like the Penan to preserve their forests. Seed believes that this work has a spiritual as well as a political dimension. Ultimate success will only come, he says, when people understand that we are nature.

John Seed
One of the qualities of the rain forest that's been noticed by a lot of very serious-minded scientists as well as by more poetic persons such as myself is that, when you spend a lot of time in rain forests, information which had previously been purely intellectual, especially information about the rain forest and about biology, becomes very personal and becomes very charged. So, for example, the information that we as a species evolved for hundreds of millions of years within these rain forests before emerging fairly recently onto the savannahs and into the open became an incredible realization.

And it was the realization that my intelligence, which up until that time had seemed so important and the mightiest thing on earth, was just the tiniest subset of the intelligence of the rain forest that had given rise to this, as well as to so much else. Every cell in our bodies is descended in an unbroken chain from the rain forest and from these vast events that the fossils and the rocks teach us about. In one sense, it was I who was fertilized by a stroke of lightning 4,500 million years ago when the first cell was created on the earth, and it was I that was the fish that learned to walk the land and felt my scales turning to feathers, and the great migrations through the ages of ice, and so on. This is my story, I feel it very personally, and it's all of our story if we are open to it and if we can lose our fascination with the sad 16,000 years of written history, which is all that we've usually identified with up until now.

*** *** ***

In Australia a few vestiges of the original rain forests have been preserved. In the Solomon Islands, the people of Paradise will keep the forests on which they depend. But in Sarawak, a tragedy continues to unfold, and Sarawak is an extreme case of what has happened throughout Southeast Asia. Unique forests have been sold off at an incredible discount; countries have mortgaged their futures to make a few people rich; beautiful hardwoods have been put to trivial uses; and the forest peoples have paid with their dignity, their livelihood and their way of life. For the Penan and many other peoples, it is probably already too late to save the situation. In other places, more sensible approaches are beginning to emerge; and, to these more hopeful perspectives, I now turn.

6

Preserving the Forest

> Tropical deforestation isn't an issue of empty nature which is being destroyed. It's also an issue of the people that live in those areas ... There will not be a solution to the deforestation of the Amazon unless there's a solution for the peoples of the forest.
> – Steve Schwartzman

Steve Schwartzman represents the Environmental Defense Fund in Washington, and he speaks for an idea whose time seems to be coming. The environment cannot be considered in abstraction from the cultures and ways of life that it sustains. Considering it abstractly can only lead to equally abstract "solutions" imposed by distant agencies of government.

This perspective does not deny that people destroy their environments. Rather, it affirms that there are also often people who have good reasons of their own to cherish and protect a given place. When environmental degradation is set up as a problem of people versus places, the solution which inevitably suggests itself is a self-defeating effort to repress the people. Looking at how people have actually lived in a given place suggests a different set of non-repressive possibilities. It reveals that ostensibly "environmental" questions have a neglected cultural dimension. When deforestation is viewed in this light, it becomes not a question of putting a fence around the forest but of fostering and protecting the cultures which give a human meaning to the forest's existence.

This chapter looks at promising solutions to the problems of deforestation raised in the previous pages. It was originally broadcast as part four of "From Commons to Catastrophe," and it begins with the struggles of Brazilian rubber tappers to protect their forest-based way of life.

David Cayley
Earlier this year, in Washington, a memorial service was held for Chico Mendez, the murdered leader of the rubber tappers' union in the Brazilian state of Acre. The fact that Chico Mendez was mourned in Washington and eulogized in the pages of *The Globe and Mail* was a tribute to a remarkable man, but also the cause he represented – the attempt of Brazilian rubber tappers to save the forest from which they make their living. Last year, some ninety rural leaders were gunned down in Brazil. Most of them died unmourned and uncelebrated in the capitals of the West, but somehow Chico Mendez captured the attention of American legislators, development bankers and environmentalists. This was partly due to his personal qualities, his modesty, his courage and the unaffected manner that let him talk with kings without losing the common touch, but it was also because he proposed a way of saving Amazonia's forests that made social and economic as well as ecological sense.

Backed by research that showed that in the Amazon, the standing forest is more productive than the cattle pastures for which it is being cut and burned, Chico Mendez and the Rubber Tappers' Council proposed the creation of so-called "extractive reserves." An extractive reserve is a large communal territory in which the traditional economy of rubber extraction, gathering of nuts, other useful food and medicinal plants, and small-scale agriculture can be practised and developed. The proposal grew out of years of struggle with the ranchers who are taking over the rubber tappers' traditional lands in Acre. Many of the rubber tappers had been there for a hundred years, descendants of the peasants who migrated there during the first great rubber boom in the late nineteenth century. They became, in effect, debt slaves to the patrons who controlled the rubber trade. In the 1960s and 1970s, assisted by the Catholic church and the Brazilian agricultural workers' union, they began to organize and establish some independence. Then they confronted invasion by ranchers, which touched off the current struggle. Susanna Hecht is a Brazilian specialist at the University of California at Los Angeles who has been close to the rubber tappers' fight.

Susanna Hecht
It's not for nothing that the Amazon is currently in the midst of a fairly grisly land war. There are a lot of problems in the land titling. That is, you have colonial land titles, land titles that were emitted in the sixteenth century, you have squatters' rights that were established in earlier periods, you have a whole bunch of basically precapitalist usufruct or use rights that pertain to the area, and then in the midst of this, you throw on a contemporary capitalist land economy.

Consequently, what happens is that you have a lot of people with either conflicting or competing claims on the same piece of land and the amount of land that is in contest right now is enormous in the Amazon. Also, in the beginning, in the 1960s, the Brazilian state basically began to emit land titles as if there were no tomorrow, not bothering to check on whether there were other existing claims. So you have had one of the largest enclosure movements occurring in the last twenty years that's probably ever existed. Fifty million hectares went out of the public domain into private hands.

The upshot of this is that there are many people who believe that they have rights to the areas that they've lived on all their lives, as anyone would, where all of a sudden other people come marching in saying, Ah, but I have the title to all of this area and you don't have any rights. What then happens is the law of the jungle, the law of the strongest — might makes right. For this reason, you're looking at very, very violent land conflicts which have characterized the region ever since the early 1960s. In part, then, the rubber tappers' movement is responding to the incursion of large-scale owners on areas to which they have traditional legal rights.

David Cayley
To defend these rights, the rubber tappers developed a method of nonviolent confrontation with the workers who were actu-

ally clearing the forest for the ranchers. One of their first antagonists was the Bordon Group, a big industrial firm which had cleared over 100,000 acres of pasture in the region of Xapuri, where Chico Mendez lived.

British filmmaker Adrian Cowell recorded one of these confrontations with the Bordon forest-clearing crew. This and subsequent scenes are quoted with permission from his film, *Murder in the Amazon*.

The rubber tappers approached through the forest, singing. Then Chico Mendez's cousin, Haimundo, addressed the Bordon crew. "Come here, comrades," he said, "don't be nervous. You are workers, like us. The ranchers' aim is to get everything. Once they destroy the natural wealth which belongs to rubber tappers, to you and to all workers, it will be wonderful for them, for then everywhere will be fenced and full of cattle. But then, how will we live? Ranch workers must be aware of this."

As a result of these confrontations, Bordon found its expansion blocked in all directions. Pastures degrade rapidly on the Amazon basin's poor soils, and Bordon could not survive without constantly cutting more forests, so the company sold its holdings and left the area.

"We've just come out of a battle with Bordon," Chico Mendez told his companions. "There are other landowners, but for the first time we've won a victory over the Bordon Group, the most powerful in the region."

At the same time as the Rubber Tappers' Council was blocking ranch expansion, it was also developing its own positive proposals for extractive reserves. The idea got an immediate favourable response from people like Steve Schwartzman of the Environmental Defence Fund in Washington.

Steve Schwartzman
The first central reason why I thought, and other people here thought, that extractive reserves were such an important idea was that this was the first new proposal for development in the Amazon to come from an Amazon community, from an Amazonian grassroots organization. Typically, development is planned the other way around. If it's not thought of in an office in Washington, it's conceived in an office in Brasilia which, with relation to a lot of the people who live in the Amazon, typically amounts to the same thing.

David Cayley
What Schwartzman and others liked about the proposal was that it addressed the ecological catastrophe that had been unfolding in Amazonia for the previous twenty years. The soils of the Amazon region are some of the worst in the world and once the forest is cleared and they are exposed, they degrade rapidly. Pastures generally become useless in less than ten years. In the neighbouring state of Rondonia, more than half the recently cleared lands are already abandoned and supporting only scrub. Extractive reserves were a possible way of maintaining livelihoods and the land base at the same time, so Steve Schwartzman went to an Acre *serengal*, a traditional rubber tapping area, to investigate further.

Steve Schwartzman
If you look at the cash income that rubber tappers make, which is very little, but then add up the other things that they do that involve small-scale agriculture that's done on a sustainable basis, small-scale livestock, collecting, hunting, fishing, depending on the area, and put a price on the parts of that which you can quantify, you see they do produce more value per hectare over even a relatively short period of time than cattle ranching and over a little bit longer time than agriculture as well. The rubber estate, the serengal, where I did my research has been occupied for about the last hundred years, and people haven't deforested more than about 5 per cent of the area. It's densely occupied. A certain amount, it's a bit difficult to tell, has been used, but there's a very large proportion where the rubber trees and the brazil nut trees are that the people don't clear for any reason. It's maintained over the long run. People live there and they have a better standard of living there than they would in the other options that are open to them.

David Cayley
In March, 1987, with Steve Schwartzman along to translate, Chico Mendez took his campaign against deforestation to the Intercontinental Hotel in Miami, where the Inter-American Development Bank was holding its annual general meeting. His particular concern was the $200 million loan which the IDB had given to Brazil to pave the road from Rondonia into Acre. The rubber tappers had seen the devastation that had followed

the paving of the road to Rondonia, and they didn't want Acre to suffer the same fate.

After a favourable reception in Miami, Chico Mendez carried on to Washington, where he lobbied the Congress to suspend its support for the Acre project. He won the ear of Senator Robert Kasten, the ranking Republican on the Senate's appropriations subcommittee, and several months later, Kasten's subcommittee followed through.

Senator Robert Kasten
This subcommittee, in a unanimous vote, without any discussion or without any objection, has taken $200 million away from the IDB. On a request of $258 million, we gave them only $58 million. Projects like the Acre project are simply not acceptable. We are not going to fund them, we're not going to pay for them, and today is the first step in terms of the Congress taking away the money.

David Cayley
The Inter-American Development Bank subsequently voted to suspend the loan. This also helped the Rubber Tappers' Council to gain a respectful hearing from the Brazilian government. But while they were enjoying national and international success, trouble was brewing in Xapuri.

A rancher called Darli Alvez, the owner of the Parana Ranch, had decided to take on Chico Mendez and had acquired the rights to Serengal Cachoeina, a 15,000-acre estate where Chico Mendez had grown up. During the summer, two prominent local rubber tappers were murdered. "We're in immediate danger," Chico Mendez told his comrades in the fall of 1988. "We're seeing people killed and there will be many more. There are dozens of names on their list of death and we must avoid this, for after death, we're useless. Living men achieve things. Corpses achieve nothing. The Parana Ranch is terrorizing the whole population of Xapuri to strike at me and the whole directorate of our workers' movement. We are on the black list."

In the midst of this terror, the Rubber Tappers' Council won another great victory. The minister of agrarian reform came to Xapuri, met with the families of the murdered men, and then signed an order creating Brazil's first extractive reserves, one of them being the area purchased by Darli Alvez, Serengal Cachoeina. The rubber tappers of Serengal Cachoeina celebrated their victory, but the order that created the reserve was also Chico Mendez's death warrant.

Cheated of his prize, Darli was determined to get even, and on December 22, 1988, Chico Mendez was shot. He had been at home, playing dominoes with the two police guards who had been assigned to him. When he left the house to shower, two gunmen killed him. The policemen fled. Federal police took over the investigation. Eventually, Darli's son, Darci, confessed to the murder, then withdrew his confession. In the end, both father and son were charged, and are now on trial. The police are also investigating the possibility that the murder was a conspiracy involving the Rural Democratic Union, the party which represents the Amazon's big ranchers.

Meanwhile, the world mourned Chico Mendez.

Steve Schwartzman
The repercussions of his death both nationally and internationally, were truly unprecedented. There was massive press coverage in Brazil. You have to put this in the context of a level of rural violence and killings in the interior of Brazil that's really just terrifying. Since 1980, by conservative estimates, about a thousand rural union leaders, small holders, Indians, peasants, church activists and so on, have been killed. Three gunslingers have faced charges. Maybe one or two of them are actually in jail, and none of the landholders who are behind these things have gone to jail, ever. There's never before been the kind of reaction that there was to Chico's death. In part, that's because there was such enormous international reaction. He was already known internationally. But it's also because he brought together constituencies that no one else had.

There's a growing environmental movement in Brazil that was beginning to know who Chico Mendez was, and the importance of the rubber tappers' movement and their proposals, before he was killed. There was also concern within church groups that have worked on human rights issues and land conflicts, within the union movements in Brazil and sectors of the urban intellectual élite, including some of the press, who, as they pay more attention to what goes on in the Amazon, have become revolted. Chico brought a lot of those people together in Brazil and they had an unprecedented reaction.

David Cayley
The death of Chico Mendez demonstrated the impact he had had: twelve extractive reserves have now been created in the Amazon. But his death also showed the vulnerability of the movement he represented. Establishment of extractive reserves is more than just a way of protecting the forest. It's also a direct challenge to the power of the ranchers who effectively rule in the Amazon. In the face of this power, says Susanna Hecht, the rubber tappers and other rural unions remain terribly exposed.

Susanna Hecht
They have international friends, but international friends didn't keep Chico Mendez alive. In fact, I left a week ago and one of the major spokesmen already had had another assassination attempt on his life. The fact is that the investigations on the attacks and murders of rubber tappers and people active in rural struggles don't go anywhere. The rubber tappers themselves don't have the money to hire a full-time lawyer. They're receiving legal assistance from some of the union officials in the south. So you're looking at a situation where they're really in a very, very precarious situation and left to face a very violent and ruthless sector of the population. As Roberto Cayal, who's head of the Rural Democratic Union, which is essentially a right-wing vigilante group and very much associated with large-scale livestock producers, said right after Chico Mendez's death: "The Amazon is ours."

David Cayley
In the wake of Chico Mendez's death, his companions gathered to celebrate mass, and they made this pledge. "I promise, before the blood of our companion, Chico Mendez, to continue his work, to show our enemies that they will never succeed in silencing the voice of the *seringueiras*. Chico Mendez, wherever you are, don't grieve that they have silenced your voice. Your ideas exist among us."

The struggle of the rubber tappers in Brazil merges economic and environmental issues, and this convergence is reproduced in the struggles of forest peoples all over the world. Ted

Macdonald has seen the same phenomenon in the Amazonian region of neighbouring Ecuador. Macdonald works for a Harvard University-based organization called Cultural Survival. He went to Ecuador as an anthropologist with a scholarly agenda of his own, but stayed to study the people's economic concerns.

Ted Macdonald
I went into the field in the mid-1970s, which was the period of extreme interest in hallucinogenic drugs and shamanism and such. I had done some preliminary work in 1972 and had identified a community which had one of the last categories of the most powerful shaman in the Upper Amazon. When I got in, I found that everybody in town was much more involved in things like cattle raising and this and that, and even the head shaman himself had hardly enough time to talk to me. So it seemed rather ridiculous to go in and do a study of something that wasn't of major importance to the local folks. It was of major importance, but there were other activities going on in the area which were definitely a priority, and how a community that had been a forest two years before was being rapidly converted to a pasture at the time startled me. I said, This has to be interesting. I wanted to know what was happening and why people were changing their economy.

David Cayley
The people in question were the Kechua Indians, and Macdonald discovered that they were basically being forced into cattle rearing. Once they had occupied a large communal territory, divided amongst the community according to clearly understood use rights. But the government saw this form of land use as nothing more than a bunch of Indians wandering aimlessly in an unmapped wilderness.

Ted Macdonald
The Kechua were told by the agrarian reform agency, which is the agency that handles most land titling, that they could very easily lose their land to alternative claimants if they didn't have title to it, and that there were two forms of title. One would be communal title, which the agrarian reform agency was extremely hesitant to give. The other was individual lots,

which they were dying to give out to everybody because it cut down the size of the community. And I think people were saying, well, goodness, we just can't keep moving farther and farther into the forest because we're beginning to bump into each other, and out of absolute necessity, groups were saying, we've got to lay claim to what's available. They were acting in a very individualistic manner because they didn't have a broader political structure to work with.

In addition, the government wanted to expand cattle production. As a result, they were making credit readily available for purchasing cattle. All one had to do was demonstrate that one had title to land and there they were, literally, with a briefcase full of money to give away to you. And if anybody flies over the area or comes into the area, a pasture is a very visible form of land use, so if someone comes in to lay claim to the land, Indians can say, well, look, we're putting it to use. It was a defensive response.

David Cayley
Cattle rearing in Ecuador ran into the same problems it had encountered in the Brazilian Amazon: the land degraded and excessively large areas were required to sustain each animal. Moreover, for the Kechuas, it was culturally alien as well.

Ted Macdonald
It was a disaster. I didn't know of any Indian in the area who made a profit. Most of them had trouble making payments on their interest. It was sort of the debt problem in miniature, twenty years earlier. Besides that, they couldn't stand the animals. They hated the size and smell of these beasts and it was not in any way part of their culture. The women, in particular, who had to tend the animals, just hated the thing. They would have done anything to get out of this business. But there were no alternatives. No one was suggesting that they get into natural forest management or agroforesty, or any of the techniques that are now regarded as much more sustainable land-use systems. This was what one did.

David Caley
Eventually, the disillusioned community began to get organized, and with political organization came a new militancy and

cultural pride. The Kechua communities, along with other native communities associated in the National Indian Federation, fought a long, drawn-out battle with successive Ecuadorian governments to re-acquire communal land rights. Some of these battles continue, but in many cases, their land tenure is now secure enough for people to begin to ask the next question: how to use these lands, if not for cattle.

Ted Macdonald
It's a very critical moment because of the rising interest in environmental issues such as sustainable development. Indians are traditionally sustainable developers. But there's also a movement towards the cash economy. As a result, they're going to have to develop some way of generating cash income for the member communities, and the hope is that they will do so in a sustainable manner. What we're trying to do is not tell them what to do, but present them with a series of alternatives, ranging from non-wood forest products to management of natural forests in a sustainable manner, to becoming involved directly in issues of conservation and tourism. Not simply being guides for some tour agency, but actually taking control over tourism in areas that are extremely fragile, that are not going to be utilized, and working to improve some of the agricultural systems so that they can generate more income.

David Cayley
In proposing these alternatives, Cultural Survival took an interesting tack. Instead of importing foreign experts, they have tried to link the Kechua with other Indian communities that have also faced the problem of devising a modernized, income-producing, but still sustainable land-use strategy. At the moment, two Kuna Indians from Panama are in the Kechua communities, sharing their experience of creating, in their fragile rain forest, a forest reserve oriented to scientific research, ecotourism and low-impact extraction. Another project at which the Kechua are looking is in the Pacazu Valley in central Peru. It's called the Central Selva Resources Management Project and it may be the first successful experiment in sustainable logging in the tropics. The project draws on the work of tropical biologist Gary Hartshorn at the La Selva re-

search station in Costa Rica. There, Hartshorn studied the role of the gaps created by treefalls in natural forest regeneration.

Gary Hartshorn
Four hundred and fifty native tree species occur in the La Selva forests. Half of those tree species require gaps in the natural forest for successful regeneration, and several of those species become big trees, and they are valuable timber. I thought that perhaps we could use this phenomenon of gap dependency as a management tool, and so I proposed creating artificial gaps in a tropical rain forest as a management technique. Many of the attempts at managing tropical forests, as we look back now at those innumerable efforts, failed because they didn't open up the canopy enough, they didn't give those gap-dependent or shade-intolerant species enough light to regenerate.

David Cayley
Gary Hartshorn went on to devise a system of logging, which drew on his discovery, a system which, in effect, mimicked natural forest regeneration.

Gary Hartshorn
The technique that I designed was to do what we call strip clearcuts. Now, in many parts of the U.S., particularly in the Pacific northwest of the U.S., maybe the Pacific southwest of Canada, clearcutting is a polemical issue. It's caused a lot of arguments between foresters, loggers, developers, environmentalists and ecologists. Just using the word *clearcuts* in the tropics has scared some people and caused kneejerk reactions. But what we do is very different from what is traditionally done west of the Rockies. We clearcut extremely narrow strips, only 30 to 40 metres wide, 100 to 150 feet wide, and a few hundred meters long. The reason the strips are so narrow is that we want to promote natural regeneration in those strip clearcuts. Remember, in natural regeneration seed dispersal is carried out primarily by birds and bats, so they've got to fly over the area, they've got to land on limbs that are overhanging the strip clearcut, or land in the young trees that are regenerating in the strip. And even though there are some valuable species that might be cut as small individuals, we're finding that they regenerate from sprouts incredibly well be-

cause we don't permit burning or cropping or farming of these strips, so we get fantastic vegetative regeneration from the stump sprouting, as well as all those hundreds of species that are pumping seeds into those strips.

David Cayley
The strip clearcut system represents a revolution in tropical American logging. Traditionally, Latin American forests have been exploited for only a few preferred species, like mahogany. This has opened large areas to settlement by driving roads through the forest, while taking out only a tiny volume of wood. The Central Selva system protects the forests by managing a small area intensively.

Gary Hartshorn
We found that the average amount of timber taken out in eastern Peru was something like 3 to 5 cubic metres per hectare, through traditional high grading. That's an incredibly small amount of timber per hectare. What we are doing is using what's commonly called appropriate technology. We use oxen to log these forests and the average volume we are taking in the Peruvian valley where we are working is 250 cubic metres per hectare. That's almost two orders of magnitude difference, and by using appropriate technology, oxen — we could even use water buffalo perhaps — to skid out the logs to a landing where it then goes on a wagon or a truck to be hauled to the sawmill, our logging costs are far less and the damage is far less.

   Even more interesting to a lot of people is that we are doing this with an Indian tribe. We helped the Indians form the first forestry co-operative in the Amazon, and they control the forests. They have the sawmill and post-preservation plant on their property, they are being trained not only in forest management and running a sawmill and doing those kinds of things, but in business administration, marketing, and so on, and they are the ones that are doing it.

David Cayley
The Central Selva project has many attractive features. It's sustainable, it preserves the full species diversity of the natural forest, and it's community controlled. It could easily become

a model for the Kechua and other forest communities around the world. But curbing tropical deforestation will depend on more than just improved logging techniques, important as these are. It will also depend on changes in the government policies that stimulated deforestation in the first place.

Theodore Panayotou is a research associate at the Harvard Institute for International Development and he represents a growing number of researchers who are beginning to point to the crucial role of misguided policies in tropical deforestation. Often, these policies do not apparently relate to the forests at all. One problem, says Dr. Panayotou, is the unwillingness of many governments to ensure that poor people enjoy secure ownership of land.

Theodore Panayotou
In a tropical environment, particularly in the uplands, it is better for soil and water to plant trees instead of annual crops, such as maize or cassava. But to plant trees, which take a long time to give you a return, you must have security of ownership. If you don't know whether you are going to have ownership when the trees bear fruit, then you will not plant trees.

Another example is that when you cultivate the land in the uplands, you need to protect the soil, to terrace it, and if you do not have security of ownership, you have neither the incentive nor the access to borrow money because you have no collateral. You have no title, you have no collateral, and therefore you cannot borrow money to invest in land improvement to protect your land from soil erosion.

Another policy which is not usually thought of as directly related to deforestation, but it's a very powerful one, is the subsidization of capital-intensive activities by governments. Capital-intensive industrialization. The result of this is that this new sector, industrial services, employs very little labour. Because they get subsidized capital, they don't employ much labour. The fact that minimum wage laws exist also discourages the employment of labour. The result is that there are a lot of people without secure land ownership, without secure jobs, and their only access to livelihood is to encroach on public property and state forests, through illegal activities such as slash and burn or poaching.

David Cayley
The forest is destroyed when people are displaced and perceive forest clearing as their only option. Often, the policies that displace people are found in the agricultural sector, where debt-burdened governments foster large monocultures of export crops, like soyabeans or cotton, in order to earn foreign exchange. This leads to consolidation of land and dispossession of small landholders, who are then forced to clear marginal forested lands. Finally, people are displaced by big government projects, like dams and irrigation systems.

Theodore Panayotou
The government, with help from some development industry, comes in and builds a dam in order to provide water to downstream farms. As a result of the building of the dam, there is flooding of a significant area in the valley. The people who live in the valley lose their land and they move to the uplands, on the slopes of the reservoir. Now they have to clear five or six times as much land to get the same productivity because land on the slope is not as fertile as the land they lost in the lowlands. When the rains come, they wash a lot of that soil into the reservoir. The reservoir loses its water-holding capacity and floods the area downstream. So the government decides that we have to raise the height of the reservoir to protect the people downstream. When the reservoir is raised, the people living on the slopes lose their land again because the reservoir now covers a larger area. They have to move further into the uplands, they have to clear ten times as much land to get the same productivity, because again it's higher up and less productive. More erosion occurs and more soil is washed into the reservoir. Again, the reservoir loses its capacity, and so forth, in a process of unsustainable development.

David Cayley
Theodore Panayotou is now able to point to cases, like the recent Dumoga irrigation project in Indonesia, where these errors have been avoided, but his hypothetical case still covers 90 per cent of the irrigation projects in Southeast Asia. They have been based on one-dimensional thinking that simply ignores unwanted side effects. Usually, the forests have paid the price. Saving what remains of the forests now depends on the

emergence of a more holistic approach that recognizes all the benefits to be derived from forests and all the costs associated with destroying them.

Theodore Panayotou believes that if the forests are to be better managed in future, they will have to be managed by communities with a material interest in their preservation. In the colonial and post-colonial era, most of the forested lands in tropical countries have been transferred from communal to state ownership. The states which appropriated these forests often lacked the knowledge, the interest and the means to protect them. Today, communities all over the world are trying to enforce or acquire communal rights to the forest, and Dr. Panayotou thinks that they are apt to exercise these rights with a lot more vision and more circumspection than governments have usually done.

Putting the community first is also at the heart of a revolution which is going on in the science of forestry. Because the poor and the landless are usually the ones who encroach on forests, the prevalent tendency in forestry until recently has been to divide and oppose the interests of forests and people. Now a movement has arisen which seeks to unite preservation and sustainable use. Bill Burch directs the Tropical Resources Institute at Yale University, and he sees this movement as partly a recovery of the past.

Bill Burch
I would suggest it's a rediscovery of traditional Asian forestry. Two thousand years ago, forestry in India was primarily a social science, and it was directed very much towards local community, towards human needs, towards applying the knowledge available at the time to increase on a sustainable basis the benefits, goods and services from a forest ecosystem. That's a long way of saying that people's needs were considered, and they directed the activity of the forester, and the forester attempted to use his or her knowledge to sustain those desired goods, benefits and services, which varied from community to community, from forest ecosystem to forest ecosystem.

David Cayley
This began to change, Burch thinks, when colonial régimes applied reductive Western models emphasizing containment

and control. He traces these models to fourteenth-century Germany, when rampant deforestation gave rise to an emphasis on the protection of forests from the people.

Bill Burch
What happened during the colonial era was that the British and the North Americans brought in the visions of fourteenth-century Germany, which was a quite different social system. This model of forestry management was restrictive, militaristic, and constraining. It was no longer proactive. It dealt primarily with commercial products, rather than the stream of goods and services, that is, snakeskins, birds' nests, all kinds of herbs, and other kinds of things that people get from the forest. These became subordinate to producing teak or other kinds of desired timber products for the British market.

And like the Germans, you walled off the forest, you got a militaristic cadre like at Dehradun, the primary forest training centre in India, which is very militaristic. They wear uniforms, guns. And that made it very difficult for them to adapt. You got movements like Chipko, the tree hugger movement in India, where people said we want indigenous species, we don't want exotic eucalyptus, we want different kinds of benefits and timber. We think the foresters are destroying the forest, not the people, and so forth and so on. The forestry professions had a hard time adapting to that, and wherever the British, French, North Americans spread their forestry message and trained people, it was in that custodial, restrictive, commercially driven kind of forestry.

David Cayley
Integrating forests with society will tend to protect them, simply because of the many benefits communities derive from the standing forest. But Bill Burch does not believe that blanket protection is either possible or desirable. The key for him is dedicating all land to its most productive use. For example, clearing lowland forests on productive soils will actually tend to protect more sensitive upland forests by concentrating food production on the lands most suited to it. As an example of the kind of integrated planning he favours, Burch points to the development of the royal forest near Prachim Buri in Thailand. Here, the people were encroaching on a protected forest, so

the government, at the instigation of the king himself, decided to develop the area.

Bill Burch

What they did was say, okay, let's look at this landscape. We will create reservoirs, we will create forests on the upland slopes, we will create national parks for wildlife and tourism development on these high lands. We'll do some timbering on the foothills of the mountains. We will create model villages where people will have electricity, schools, potable water. We will develop these reservoirs for irrigation of these formerly royal forest lands and we will have the foresters work with the people to replant within the fields, within the growth of cassava and other kinds of crops.

One of my former students has been one of the active foresters in the design and development of this, and when I was over there a year or so ago, we went out and talked to some people. We went to where foresters were making decisions about the planting of trees based on how it would affect the quality of honey, because that's what the local farmers wanted. The honey from their bees would be marketable in Bangkok and elsewhere.

The design of the project really follows the model of what we think is the ideal kind of mosaic of uses. Where you should have forests, you have forests, where you should have parks, you have parks. All of those are reverberating in terms of economic benefits. There's no talk of insurgency and so on in that province any more, but it's that mixture of benefits where the local population are participants in the sustainability of that mix. The foresters are like community developers bringing knowledge about ecosystems and plants and trees, but in the service of the needs of the local population. It's an amazing thing to see.

David Cayley

The Prachim Buri project offers a model in which conservation is integrated with production. It's an approach that recognizes that tropical forests can only be saved if the poor and landless people who now encroach on them have alternatives.

Environmentalists often express the conviction that tropical deforestation is driven by international forces, ranging from consumption of tropical forest products, to the lending practices of the World Bank, to the debt crisis. This belief has the comforting corollary that it is within the power of the rich, industrial nations of the temperate zones to end tropical deforestation by taking the appropriate actions. There is certainly some truth to this. Many destructive projects that depend on foreign financing wouldn't take place if that financing weren't available. But Susanna Hecht also thinks that it overstates international influence.

Susanna Hecht
It isn't an imperial world, and people in the U.S. and in Canada are not going to have a lot to say about what goes on. That is, it's not all run by the World Bank. Brazil is the world's eighth largest economy, so while it has a large debt, it doesn't necessarily adhere to what's dictated by the international bankers. The internal dynamics of these countries are often far more important than the international context in which they're inserted, at least in reference to deforestation.

David Cayley
There is no necessary dichotomy between local and international action, of course. Stopping the international financing of destructive projects does help to stem deforestation, and changes in the international climate of opinion have created pressures for change in tropical countries. Chico Mendez didn't go to Miami and Washington for nothing, as I'm sure Susanna Hecht would recognize, so it both is and isn't an imperial world. But it is certainly true that forest conservation will depend in a critical way on popular initiatives in tropical countries, like the ones I described earlier on. Ted Macdonald of Cultural Survival has worked with the organizations that represent the indigenous peoples of Latin America, and he sees an historic opportunity for these organizations to join forces with the growing environmental movements which exist both in Latin America and North America.

Ted Macdonald
The indigenous organizations should really begin to realize that the moment is ripe for them to make their point. The world is concerned about the tropical forest. How long the

world stays concerned remains to be seen. How long indigenous people are concerned with the tropical forest does not remain to be seen. They will retain that interest, provided they have the opportunity to do so.

Consequently, the opportunity for convergence between those who are concerned with environmental issues and those who are concerned with proprietary rights first and foremost, in other words, the Indians themselves, is really unique, and the potential for coming together and working effectively is there. Both groups could realize their goals by working together. However, it's going to require more understanding on the part of environmental organizations, environmental planners, development workers, and others, of what it is that Indians are really concerned with in a broader sense, that they have these concerns over land, that they've had a lot of negative experience with development, that they tend to view the standard concept of environmentalism as protection of fragile lands against human occupation and utilization.

And so, in many cases they regard environmental organizations or environmentalists, whatever, in much the same light as they would regard development people. They are folks who have a plan for the use of their land, which doesn't necessarily include them in the planning of that plan. And once the organizations, both the sensitive development groups and the environmental organizations, begin to realize that there is much greater potential for convergence of interests, once they're willing to look at some of the broader concerns of indigenous organizations, we will really begin to see some long-range impact.

*** *** ***

Ted Macdonald's point can be enlarged to apply to other communities as well. Forests are destroyed by people who have no material interest in their preservation, whether they are rich ranchers, poor peasants or remote governments. They will be preserved only when the forest and the community are integrated. Tropical deforestation is a big problem, but it may have a lot of little solutions.

IV
The Age of Ecology

7

A Managed Planet

During the late 1980's, calls seemed to issue from every quarter for planetary management, world strategies, and global thinking. The image of the whole earth was instantly recognized — as a religious icon, as a corporate logo, or as a come-on for running shoes and biodegradable detergents. Television viewers were enjoined to "save the planet." People appeared in T-shirts announcing them as crew members of spaceship earth.

One of the signs by which I noticed this new global approach was people's habitual use of the term *we* — as in, "we are destroying the rainforest" or "we are making the earth uninhabitable." I pondered about who this *we* might be. Plainly, these things are not done by humanity in general, but by specific groups of people. Why, I wondered, attribute the actions of real men and women in real historical circumstances to an imaginary collective subject? Then it began to occur to me that the environmentalist's *we* might not be intended as a description, but as a heuristic device — a way of guiding people from local identities into a new planetary and pan-human identity. The use of *we* invokes a reality that is desired, rather than describing one that already exists.

This new planetary discourse is intoxicating, as its popularity shows, but it is also disturbing. Is the planet a mental, moral, or physical space that we can actually occupy? Until now, the peoples of the earth have interpreted the time and space they occupied according to the canons and conventions of their own cultures. Our history belongs to us as residents of a certain place, citizens of a certain country, members of a certain religious community or whatever. Is our biological identity as a species sufficient to unify us? Can politics or citizenship be imagined on a global level?

The unifying theme of current ecological discourses invoking this new planetary reality is usually survival. For humanity to survive, it is said, the earth must be brought under coordinated management. In this sense, ecology presents itself as the new transcultural narrative underlying a planetary identity. What unifies us is the earth itself and its interlocking ecosystems. This unity is very different from that foreseen by Christianity or subsequent Western humanisms. It rests not on moral or religious grounds but on a scientific and technological basis. And from this fact hang a series of questions about whether ecological universalism can be anything but unification at the lowest common denominator.

With these questions in mind, I decided that I wanted to try to create a series of programmes that would explore the many meanings of ecology. It seemed to me that what might be called mainstream, or managerial environmentalism, was rapidly hardening into new orthodoxy, a new common sense. I wanted to do something that would be based on open questions, rather than foregone conclusions, something that would invite perplexity — as well as alarm. Having thought myself for twenty years as an environmentalist, I now found myself on the sidelines muttering, like the woman in Eliot's *J. Alfred Prufrock*, "That is not what I meant at all. That is not it, at all." The rhetoric of survival suggested to me only what ecologist Stuart Hill calls "humanity in retreat" — not a vision of a flourishing human community but of people grimly clinging to their branch by their fingernails. The new vogue for green consumerism seemed to throw into the shadows more awkward questions about how to limit the power of market relations themselves. As environmentalism joined the mainstream, it seemed to be losing its ability to imagine a different society and focusing only on those defensive transformations which would sustain the existing way of life. I wanted to question the new common sense and expose its hidden assumptions. The result was "The Age of Ecology," a collection of interviews broadcast on CBC in June, 1990.

The first speaker in the series was Wolfgang Sachs, a German thinker whom I had the good fortune to meet through Ivan Illich at the time I was beginning to wonder about the shadow side of ecology. He, too, was critical of the direction mainstream environmentalism was taking, and his writings helped me to understand what bothered me about the new environmentalist discourse I was hearing all around me. Sachs grew up in the German Green movement of the 1970s. He worked on alternative energy policies for Germany as part of the research group on energy and society at the Technical University of Berlin. He wrote a book on the life of the automobile, now being translated into English, and then, in the early eighties, edited a journal called *Development*, published in Rome. At the time we met, he was working with Illich and teaching for part of the year at Penn State University. I spoke with Sachs at State College, Pennsylvania in the fall of 1989.

Wolfgang Sachs
What bothers me today is that too many people talk too easily about the "environment." What was once a call for new public virtues is now about to be turned into a call for a new set of managerial strategies. When one sees how the World Bank begins to move into environment, when one sees how the experts of yesterday, the industrialists of yesterday, the planners of yesterday, without much hesitation, move into the field of the environment and declare themselves the caretakers of the world's environment, the suspicion grows that the experts of these institutions have now found a new arena to prove their own indispensability, and all of that in the name of ecology and the survival of the planet.

Development has been an intervention in the history of many countries in order to boost the GNP. Now, with the alarm that the survival of the planet is in danger, we slowly move into a situation where there is no limit to intervention any more, because can you imagine any better justification for large-scale interventions in people's lives than the survival of the planet? For me, that means I have to step back and to ask myself, What is happening here, what are the new distinctions we have got to make?

David Cayley
As Sachs began to draw these distinctions, he noticed that environmentalism had always contained contradictory impulses, that these contradictions, in a sense, constituted the science of ecology.

Wolfgang Sachs
The very term *ecology* implies an ambivalence. Ecology can be a demonstration on the street and it can be a computer modelling. Ecology can be political action as much as a strict and sober academic science. On the one hand, it is a movement that attempts to put science and rationality into its place, a movement that in its deepest motives, has an anti-modernist gesture to it. On the other hand, it is also a movement that claims to call for the better science, namely it takes reference to ecology, which is an established science, and criticizes today's rationality and science in the name of ecology. So it takes resort to a modern science in order to push anti-modern aspirations. This hybrid character of ecology, however, is the secret of its success. Ecology combines two tendencies: it combines a call for science, which is the religion of modern society, with a call for less science and rationality, which is its anti-modern heritage. So it combines modernism and anti-modernism and, if you want, it is the first anti-modernist, conservative movement that attempts to fight its enemy with its own means.

David Cayley
So long as environmentalism remained an opposition movement, the contradictions contained in the word *ecology* provided the movement with a sort of hybrid vigour. It could straddle both sides of the issue and get away with it. But now that environmentalism has advanced at least to the vestibules of power and influence, this contradiction is increasingly exposed. For Sachs, this means that the very different meanings of environmentalism now have to be distinguished and two very different political options set against each other.

Wolfgang Sachs
Take the problem of fuel consumption in the automobile and the pollution caused by the automobile. There are two distinct ways to deal with these problems. One approach is to say

nothing can be done about the way we use automobiles: they have become second nature. So then you are left with trying to increase the efficiency of those automobiles. You will build them to run more kilometres with the same gallon, and the result will be fuel-efficient cars and fuel-efficient engines. The second approach would be to ask yourself, What do you want from an automobile and how many automobiles do you want? You would try to put a brake on the overall number of automobiles, and in particular the performance of automobiles. You would begin to talk about a low-speed car because speed is the single most important factor in energy consumption and pollution. You would, for instance, try to invent an automobile which, because of its construction, cannot run faster than fifty miles an hour and an automobile which then has its point of highest internal efficiency at, let's say, fifteen or twenty miles an hour.

When you do that, all of your parameters for the construction of automobiles have changed. You need much less sophisticated technology. You don't have so many problems with weight any more, you don't have to have all the safety precautions that are built into these automobiles, you have less consumption of land for streets, you have a lower number of mortal accidents, you have lower energy consumption, you have lower pollution.

My point is that I would like to reconsider our use of automobiles and what we expect from automobiles, and I would like to call for, if you want, a slower society. I would like to call not for the wholesale abolition of the automobile, but for a more intelligent automobile, for a moderate motorization. It's an approach that asks what do we want, what do we aspire to, what do we want to work for, how do we want to live, whereas the second approach assumes that we cannot change the way we live. All that we can do is to try to make more out of less, to make it more efficient, to manage it better, to streamline it. So where the second approach, or the management approach, is going is a more monitored and well-tuned society in the name of ecology.

David Cayley
This new society preaches what Sachs calls "the gospel of global efficiency." Instead of choosing freedom through self

limitation, it is choosing high consumption under ecological surveillance. The movement that Sachs hoped would reduce the grip of constant economic calculation on our lives now threatens to increase it.

Wolfgang Sachs
Take, for instance, one word which has made a tremendous career, the word "risk." Today, everybody talks about risk, risk management, and risk precaution. The word gives life to quite a number of new departments in governments and also universities.

I remember very well that in the 1970s when it came to nuclear power and chemical plants, we talked about dangers. We talked about dangers, threats or possibly hazards. Now look at the language. If you have a child who goes to school every day, and on the way to school there's a pit, then the pit represents a danger. So what do you do? You remove the danger by putting a fence around the pit or putting a board over it. The moment you say that the pit is a risk, something happens: a new attitude emerges. You think, maybe I shouldn't put a board over the pit because the board costs such and such. The risk that the child falls in may not be that high. You begin to calculate, to weigh one factor against the other.

The more we talk about "risks" and not about "dangers" or "threats" to our lives, the more we fall into the language of cost and benefit, of calculating, and of weighing one cost against the other, the more we imply management, monitoring, risk control, or continuous supervision. Something happens in the language, and in the language, you can follow this trend toward environmental managerialism.

David Cayley
One of the reasons Sachs fears this new managerialism is that he sees it as a homogenizing force: the astronaut's gaze perceives no boundaries, only the flowing forces of planetary ecology.

Wolfgang Sachs
Under the eyes of planetary management, it is as if there are no differences any more on the globe. Nations fade away, interests recede into the background, cultures are used only in

an ornamental way. There is no "other" camp on this world. The world somehow merges into one. Since the Enlightenment, people have called for a unity of humankind, but this was a moral postulate. We would strive to overcome war and violence, and people would strive to unify mankind under the governance of reason. But now, the unity of the planet, is occurring as the result of fear. It's the result of a menace, a threat, the final catastrophe. This menace, in our perception, creates a homogenous global space where differences between cultures, between men and women, between nations, between top and bottom, don't matter any more.

David Cayley
To verify what Sachs says, you have only to turn on your television. You'll soon hear a lament about what *we* are doing to our environment. But this *we*, for Sachs, is a depoliticized and depoliticizing category, a night in which all cows are black, as Hegel says. It omits distinctions of class, country, or social system and reduces everything to a biological, dead level.

Wolfgang Sachs
In the 1970s we fought against nuclear power plants and against new highways, as citizens who wanted to have a different life, as citizens who had a different notion of what the good life is about. Today, I notice that the citizen doesn't exist any more. When you look into a *Newsweek* report, you don't find the citizen there, you find the human species there. So human beings are not called citizens in the most recent environmentalist discourse, they are called species, and the problem they are facing is not what we used to call "quality of life," but the problem of survival. You don't have societies or communities any more, now you have populations. Consequently, I think that this language brings along a biological reductionism which again has the function of eliminating many things which make us human, namely that we have different aspirations, that we have different notions of the good life. And I see this new discourse as another sign that under the banner of ecology we are moving into a new phase of making the world more uniform.

David Cayley
The elimination of citizens is also implicitly the elimination of a civic space in which citizens can act. Biological language, in effect, drives out political language.

Wolfgang Sachs
It is a grand operation to render politics irrelevant. Imagine that the World Bank or some other international institution adopts this language. Talking about species, survival, populations does not give room for a political or moral discourse. You no longer ask, How do we want to live, how can we responsibly live, what do we want to produce, how do we want to do it, how do we want to arrange our lives, what are our aspirations in life? What happens is that all kinds of political distinctions evaporate, and moral argument remains without a grip.

And that, of course, is technocracy's oldest dream: to make political arguments and political distinctions disappear. Technocrats would like to reduce everything to the functional requirements, and the best basis of course, is a biological basis because, in the end, we want to survive. Once you are able to reduce a global problematic to a biological problematic of species survival, there is no political or cultural argument that can disturb the actions of technocracy any more.

David Cayley
Sachs obviously doesn't depreciate survival as such. He worries about the consequences of its becoming a reason of state and he is astounded at the irony that the wealthiest societies in history can find no worthier reason for being.

Wolfgang Sachs
There has never been a society in history for which survival would have been a prominent objective. To secure survival was a banality, something which went along with whatever greater achievements a society wanted to aspire to. Now, we have the paradoxical situation today, that at the very moment where we have amassed riches like never in our history, experts from all four corners call upon us and call upon our governments to put survival first. Now, I ask myself, what is happening here? It is important to recognize that the call for

survival assumes that in the future we will always be moving along the edge of the abyss. Traditional societies knew limits to their production and consumption. In various ways, they stayed away from the edge of the abyss. Through industrialization and the recent upsurge in industrial production, we have been pushing the limits so far that now water and soil and air have become scarce goods. In this situation, gearing society toward survival implies that we are incapable of stepping back from the edge of the abyss. It implies that we have to set up institutions, to find experts, and to transform governments in order to steer a precarious course along the edge of the abyss. Therefore, I mistrust these calls for securing survival, because they put survival first. I want a society where a good life comes first and putting a good life first, means stepping back from the edge of the abyss in order to be in the position where survival does not have to be the governing principle of social politics.

David Cayley
Survival, to Sachs, is a code for carrying on a commodity-intensive way of life under the surveillance of "ecocrats." The alternative is what Ivan Illich once called "conviviality," a more austere life, in which people would depend less on technique and more on one another. Both approaches agree on the need to do something about pressing problems, like excessive carbon dioxide emissions, but they disagree on how and even more on why to do it.

Wolfgang Sachs
It seems obvious that emissions of CO_2 are excessive. What common sense demands is that industrialized nations bring down CO_2 emissions. However, the kind of colour which the discussion takes on is that we have to fund lots of research in order to understand better all the atmospheric feedback, the feedback cycles in the atmosphere, the earth and atmospheric system, meteorology, the formation of clouds, the impact of oceans, and so forth. And to understand better means to check out the responsiveness of nature. The hidden intention is to go to the limit, to see how far can we ride the tiger.

For instance, in Germany, the minister of research has set up a programme to look into the possible effects of rising sea levels on Germany and the German coasts. So they are already spending money to make a greenhouse-related event manageable. They take for granted that we cannot bring down our level of greenhouse emissions very much. Now, I agree with those environmental managers or ecocrats who say that we have to bring down our greenhouse gases, but I think that we have to do it from a position of ignorance. We have to say two things: first, we are not able to understand all the complex mechanisms which govern the atmosphere of the planet, and second, we have to behave prudently. That means we have to keep back far away from the edge of the abyss, and that, in conclusion, means we have to bring down our emissions of CO_2 radically, in particular through restructuring our way of life, our way of producing.

David Cayley
To get to your preferred approach of limitation, of stopping short well before you come to the abyss so you don't have to then manage your careering along the edge of it forever, implies that there's a political society which can make these decisions. And I'm wondering finally if that political society any longer exists.

Wolfgang Sachs
I don't know. I can only have the hope, but I don't have the expectation. However, I would submit that in that sense history is on my side, because during the last fifteen years, all kinds of surprising things have happened. Think of Poland, think of the Soviet Union. History consists of surprises. So for that reason, I do not know what will happen. Neither do I know what the best strategy could be to make something happen. History is not an arena where strategies are played out. That means that even if I am fully aware that today's political situation is not very inviting when it comes to defining and appreciating limits to growth, I try to practise something you could call selective simplicity. Not to do some things which everybody is supposed to do, not to have a television or not to have a car, to decouple from what is considered an average consumer today. That's already a step in the right direction. Then, of course, as an intellectual or as somebody who tries to be politically active, I try to advance ideas, to advance a lan-

guage, to advance a perspective that makes more visible a politics of self-limitation. I do not know if this approach will be effective. I only would like to be there when history comes around with some new surprises.

David Cayley
Wolfgang Sachs questions the implications of replacing the traditional language of politics with the language of biological science, of transforming citizens into species, nations into populations, the bounded spaces of earth into the swirling systems of planetary ecology, and survival into a reason of state. His scepticism about ecology as a political guide echoes the earlier work of historian Donald Worster, the author of *Nature's Economy: A History of Ecological Ideas*, published in 1977. Worster, now at the University of Kansas, began his research at the end of the 1960s when the age of ecology was first heralded. It was then that ecology achieved the unique status of a science that was, at the same time, the banner of a popular movement. There were no physics parties or sociology parties, but there were ecology parties springing up all over the Western world. People entirely innocent of the academic science of ecology began to call themselves "ecologists."

"Is ecology a phase of science," asked the distinguished ecologist Paul Sears in an essay he published in 1964, "or is it an instrument for the long-run welfare of mankind?" Donald Worster wondered at such claims. "Like a stranger who has just blown into town," he wrote, "ecology seems a presence without a past." He decided to investigate.

Donald Worster
What struck me forcibly about twenty years ago was that there was a new science on the horizon, at least on the popular scene, the science of ecology. News magazines were talking a great deal about ecology, and ecologists were appearing on their covers. This new science was making quite an impression. It was being hailed by many people as a new oracle. It was the authority finally needed, the guide to get us out of the environmental crisis. It would furnish Truth, with a capital *T*. Others were beginning to argue that it provided a basis for a radically new world view, a new ethic, even a new religion. People were coining new words with "eco" attached as a kind

of prefix. Eco-philosophy, eco-feminism, eco-cities, that sort of thing. Well, I set about to examine the history of ecology. To find out what its employment record had been, to get some sense of its CV, if you like. I was interested in how science has shaped our perception of nature, over time, in order to understand where it would take us in the future.

David Cayley
What struck Worster was what he called the moral ambivalence of ecology, its contradictory character. He called the two contradictory tendencies the arcadian and the imperialist. The imperialist side could be traced back to Francis Bacon and his vision of science as the subjugation of nature, "the effecting of all things possible," in Bacon's resonant phrase. The arcadian tendency was embodied in the Romantic movement, in Goethe's vision of an ethical science, in Henry David Thoreau's wonderful description of his scientific studies as "nature looking at nature." He also noticed the metaphorical character of ecology, the way in which it reflected the attitudes of the surrounding society. Both ecology as metaphor and ecology as moral ambivalence were clearly displayed in the work of the nineteenth century's greatest ecologist, Charles Darwin.

Donald Worster
Charles Darwin is the clearest example I think we have of how a scientist working with reasons, facts, and hypotheses nonetheless reflects the society and the culture of which he is a part. On one hand, Darwin put at the very core of his science the idea of a struggle for survival in the natural world, a fiercely competitive world that clearly reflected the nineteenth-century English society in which he was living, the society of laissez-faire capitalism, industrialization, growing poverty, and urban social problems. He was aware of those things, and they affected the way in which he saw the natural world. When he looked at the natural world, he saw the same kinds of social forces that were at work in Victorian England. On the other hand, Darwin maintained a kind of vision of order and harmony in the natural world, the beauty of the whole, and I think he took that vision mainly from the Romantic poets, artists, naturalists and philosophers of the early nineteenth century

who formed a kind of counter movement to that industrial revolution, laissez-faire capitalism.

So Darwin was a man whose most interesting insights came from this contradiction, from the way in which those two tendencies in his thought worked together to create the foundations of modern ecology. And so, subsequent scientists can take both sides from Darwin and they can study ecological adaptations, the harmony of the natural world, or they can emphasize competitive exclusion, struggle for survival, the law of tooth and fang, individualism, and the subsequent history of ecology is really a debate that goes on between those two poles of thought. Is nature essentially a co-operative balanced whole or is it a world of chaos, struggle, bloodshed, murder?

David Cayley
The ambivalence of ecology has taken many forms. One of these forms has been the debate between a mechanistic approach, which has been characteristic of science since its origins, and an organicist and vitalist approach, which looks for some living principle in the natural world.

Donald Worster
That contradiction goes back well before Darwin, into the seventeeth and eighteenth centuries. Scientists mixed their metaphors a good deal then. Is nature a living whole, an organism? Is there a kind of breathing soul and spirit of the natural world that in effect makes all of nature a single organism? There were those in the seventeenth century who were arguing yes. At the same time, people were developing a quite contradictory model of the natural world, a metaphor of the machine, that nature is essentially a kind of contrived mechanism, springs, bolts, levers working, wheels turning, all of that sort of thing. But again, the philosophical implications in those two almost diametrically opposed world views are not well worked out until the later part of the nineteenth century. They lie there together, as they probably still do in popular literature and thinking or even in a lot of scientific thought. But by the late nineteenth century, there is clearly a debate, a very conscious debate going on about those two sets of metaphors and the world views that they're a part of, the

organismic and the mechanistic, and ecology is very much wrapped up in that.

David Cayley
In the early part of the twentieth century, the organicist approach clearly held the upper hand. Its leading exponent was the American ecologist, Frederic Clements.

Donald Worster
Clements was a Nebraskan, an ecologist at the University of Nebraska, who founded the first North American school of ecology, often called the Climax School. Clements's argument about nature was essentially that it goes through a series of stages, or what he called the succession, that leads finally to a climax stage, a kind of fully mature, settled stage in which the plants and animals are all in fairly perfect balance and they endure, they're stable. At that point, nature has in effect evolved into a kind of single organism. He described the prairies of North America as an entity that was so closely integrated and so harmonious in its workings that it looked like a kind of organism, perhaps not as complicated or as closely integrated as a buffalo or prong-horned antelope, but still having organismic qualities to it, and those organismic qualities in effect allowed it to continue, gave it a life of its own and permanence.

David Cayley
Clements's theory has endured in the popular mind, but as science it began to be superseded in the 1940s. Ecology veered back hard towards a more reductive or mechanistic and more easily quantified approach, and a new synthesis emerged.

Donald Worster
By the 1960s, ecology had dropped Clements and was now basing its ideas of the natural order on physics and on systems theory, and the word that replaces *climax* is *ecosystem*. It's a word coined by an English scientist, Arthur Tansley, in the 1930s, but it didn't really catch on in this country until after World War II. The ecosystem includes both plants and animals, but also the inorganic parts of the environment that Clements basically left out — the soils, the geochemical cycles

in the environment — to get a model of nature that is essentially based on the flow of energy from the sun, through the plants, on up through the animals, recycling itself constantly, the matter constantly recycling, energy being passed up the food chain and finally lost through the processes of entropy in the black hole of the universe.

This theory is mainly associated with Eugene Odum at the University of Georgia, whose textbook, *The Fundamentals of Ecology*, was the dominant one by the early 1970s and remained so throughout the decade. It's the form of ecology that most of us know. Over the last twenty years, it's been the one that has been in the news most. You talk about ecosystems a great deal, damaging of ecosystems. If you read any news reports on the oil spill in Alaska recently, they all talk about the ecosystem and what oil is doing to damage the ecosystem. That's all out of the new ecology of the 1950s to 1970s.

David Cayley
What's notable about the metaphorical expression of the new ecology?

Donald Worster
It's a combination of many, many ideas and metaphors again, some of them derived from physics and energy flow, but what's most interesting to me about it is the way it takes the language of economics and embeds it into the textbooks. Odum uses this language to describe his ecosystem, and most everyone else who's followed the ecosystem model also uses it. These ecologists see the ecosystem as being divided into producers and consumers, and what is flowing through this ecosystem is the currency of energy. So nature has become very explicitly an economy, and one that looks a great deal to me like a modern industrial consumer society, with producers and consumers all organized and circulating the commodities of the shopping malls. It's a kind of a well-run factory that nature manages for maximum productivity. They begin to use words like productivity. Economic efficiency is applied, only they call it ecological efficiency. The production of biomass is the standard by which the ecosystem is measured. How much biomass does an ecosystem produce? On top of the language of physics,

energy flows and systems theory, we've got grafted on a kind of economics of nature.

David Cayley
At the time that Donald Worster published *Nature's Economy*, he could still describe the ecology of the ecosystem as "the new ecology." It remains the prevalent popular understanding. But when Worster went back to the ecology textbooks recently to bring his history up to date, he found that, as science, it too had been superseded.

Donald Worster
The ecosystem as an idea has dropped out of the index of many contemporary ecology textbooks. What ecologists now see is not a pattern of order, but chaos, and in fact there are some ecologists today who are very much a part of the new science of chaos. When they look at nature, what they see is instability, disorder, a shifting world of upheaval and change that has no direction to it.

Clements's nature had a direction — the climax theory. That was the end point. In effect, the ecosystem had an end point, it had a direction that nature was evolving toward. But in the most recent ecology, I don't see any direction. There's a loss of confidence in any concept of order. What ecologists find when they look at an acre of land is constant change going back thousands and thousands of years. Ecologists have become historians, and they're finding very little in the way of any coherent model. If you look at the Great Plains of North America as an example and you go back a few million years, we go through forest, we go through seas, we go through grasslands. When does it end? What's the pattern here?

Clements was aware of this, Eugene Odum's certainly aware of this, Darwin was aware of it. These people all invented, in a sense, the fact that nature has a past, a history. We've all been historians of nature for a long time, but it's become far more pronounced as a tendency in recent times, with the outcome that there is less confidence that there is any coherence to any of this. It's just shifting patterns. Plants come and go, animals come and go, the climate changes regularly, nothing is predictable, the future won't be anything like the present, and so on.

David Cayley

Donald Worster began his research in the history of ecology in order to assess the claims being made for it as a potential guide. He found a kaleidoscope of images and ideas drawn from current social practice, a persistent moral ambivalence, and a wavering and uncertain image of nature — nothing, in other words, which could furnish a basis for moral decisions.

Donald Worster

Right now, if you're a policy maker and you call up an ecologist who's been reading some of the new textbooks and you ask the question, Well, what does your ecology tell us to do? What do you want us not to do?, the answers are very troubled, uncertain. If the world around us is as chaotic as some of the recent textbooks say and so full of change and upheaval, what can ecology tell us to do or refrain from doing? What does it mean to damage nature? How do we even know we're destroying nature if nature has such a troubled history? It puts the policy maker in a very difficult position if he or she is turning to science as the authority, the oracle, today.

In the eighteenth century there was no question that there was an order designed by God. Darwin had no doubt in his mind that evolution finally led to order and harmony, and he knew it when he saw it. Clements was certain that there was a climax state of vegetation that the white man had disrupted and destroyed, creating the dustbowl of the 1930s. Eugene Odum was sure that there was something called an ecosystem that could be disrupted or unbalanced, damaged in some fashion or other.

All those preceding concepts of order have fallen away. A new one may be on the horizon at any time, but right now I don't see one, as an historian looking at what's been going on in ecology for the last several years. So we're reduced to talking about nature in clearly anthropocentic terms. That is, the damage we're doing is not to nature but to ourselves, or it undermines the sustainability of our economy or our society, or it threatens human health in some fashion or other. But that's about all we have in the way of basis for policy. Maybe that's enough, maybe that's all we'll ever have, but it's rather different, I think, from what people thought we were heading toward twenty years ago, around the time of the first Earth Day.

David Cayley
The failure of ecology as an oracle leaves responsibility right where it always was, in any case, with the society that invented science in the first place. Science can show us no definitive image of nature on which to base our judgements. But for Worster, that doesn't mean that we should abandon science or moral judgements, just notice the difference.

Donald Worster
My view is that we shouldn't throw science out. It is clearly, however, shifting ground. To build a world view or an ethic or religion, if you like, on the science of ecology is like building a house on a floodplain. Sooner or later, a lot of water's going to come down that stream and wash you away. I think historians are inevitably sceptical and relativistic when they think about science as an oracle. It seems less reliable when we look at its past, and that's one of the outcomes of my own research in the history of science. But I remain committed to the idea of an order of nature, I remain committed to the idea that we don't simply talk about the damage we're doing to this planet in anthropocentric terms, and it seems to me that we have to get our heads together from all disciplines and ways of thinking to discover what that order is. We need artists involved, we need poets, we need historians, we need philosophers. We need reason. Scientists are going to be a part of that, but they're not going to give us the final and decisive answers that will solve all our policy questions and provide us a basis for morality and ethics.

*** *** ***

Both Donald Worster and Wolfgang Sachs, in their different ways, raise questions about ecology as a guide and motive for moral action. Worster has contemplated the shifting ground of ecological thought as a historian and concluded that our views of nature are never free of prejudice. Nature can be an intricate clockwork revealing and glorifying its Designer, as it appears to eighteenth-century eyes, or a magic shuttle weaving an ever-changing pattern in the void, as it appears to the contemporary science of chaos, but it is always an image of the society which authors the theory in question. This is easier to perceive in the mirror of the past, where people's ways of speaking no longer seem to us to be common sense. It is plain enough in retrospect that Darwin's nature is also Darwin's England. Our own ways of speaking — of nature as a system, for example — seem natural, and it is easy for us to overlook the fact that they are also projections of contemporary social and technological realities. But nature cannot appear to us as a cybernetic system until we have first formed the idea of a device which adjusts its state to its last calculation. Worster's work brings us up against this relativity in our conceptions of nature and suggests that, because ecology is as fallible as any other human project, we cannot flee the burden of political decisions for the security of science.

Wolfgang Sachs fears that this avoidance of decisions may be happening nonetheless. He foresees that, by seeking a mandate for social decisions in the biological imperatives of ecology, societies will bypass the question of what is good in favour of the question of what nature will tolerate. As an alternative, he proposes a politics of self-limitation which preserves freedom by setting conservative limits to economic activity and, thereby, eliminates the need for constant calculation and surveillance.

Neither thinker wants to allow ecology as "oracle" to obscure the moral and political dimensions of the environmental crisis. Science was a large part of what got us into this mess. We cannot trust the clairvoyance of some new science to get us out of it.

8

One-Eyed Science:
A Conversation with Vandana Shiva

Environmentalism in North America has often been portrayed by its opponents as a kind of middle-class indulgence, a concern only the well-to-do can afford. The charge is unfair, but it does reflect the fact that for most North Americans, environment is an abstract category. We may worry about topsoil loss, but we don't depend on a particular soil to subsist. If the avocados from California don't look good today, there are always the melons from Israel or the kiwi fruit from New Zealand. Things are otherwise in the countries of the south, whose soils are often the source of our luxuries. There, environmentalism didn't begin as what we'd call environmentalism at all. It began with people defending their own subsistence and, therefore, defending the environments in which they subsisted. One such case was the Chipko movement, which appeared as a protest against deforestation in the early 1970s in the Himalayan region of north-eastern India. It was a movement of village women who adopted the tactic of embracing the trees, which is what *chipko* means, as a last-ditch defence of their own safety and subsistence. The Chipko movement is one of the subjects taken up by Indian writer Vandana Shiva in a book called *Staying Alive: Women, Ecology and Development*. Vandana Shiva lived in Canada in the 1970s, taking a PhD in quantum physics from the University of Western Ontario. She then returned to her native Dehradun, in the same part of India that Chipko arose, where she established the Foundation for Science, Technology and Natural Resource Policy. Chipko was then at its peak and she was quickly drawn to activism as well as scholarship. Recently, she taught for a term at Mount Holyoke College in Massachusetts. I met with her there and recorded the following interview. We spoke first about Chipko; but since that material overlaps an earlier interview for "Citizens at the Summit," I have begun this excerpt from her discussion of the Green Revolution.

The Green Revolution is the name that was given to the introduction in the Punjab district of a new agriculture based on high-yielding hybrid strains of cereal grain developed by the Rockefeller Foundation. For Vandana Shiva, it perfectly illustrates the pitfalls of development as a way of receiving the world.

Vandana Shiva
There are two shifts that took place with the green revolution in agriculture that aren't talked about when the miracles about the great revolution are discussed.

It robbed the farmer of his or her mind. No more was the farmer a thinking being. He or she was from now onwards a passive receptor of external ideas coming from four or five centres of international agricultural research. And the same thing happened to the land. All agriculture, especially in the Third World, had been a fully internal input system. Trees were linked to livestock, linked to soil, reproducing each of these systems through recycling, year after year, needing nothing from outside, in fact producing surpluses that could feed other organized forms of society outside agriculture. Livestock, of course, disappeared as a part of farming with the green revolution because tractorization totally displaced the need for animal power, and trees disappeared with the green revolution because now it was just crops. Production had to be maximized, and the soil from now onwards was just a passive receptor of inputs to be purchased on markets, and the inputs were seed, which until then had been a product of that same soil and went back to that same soil and produced grain as well as seed.

Consequently, the seed was more than one entity until the green revolution. It was both the food for human beings as well as the source for its own reproduction. The green revolution created an ontological split in the nature of the seed. The seed was now no more grain and the grain now could no more be seed. Seed had to be purchased on the market, and the seeds that had to be purchased on the market were engineered in a certain way so that the internal input cycles of the seed were not relevant. In fact, the seed was manufactured to make it free of these internal input cycles and to make it dependent on external inputs of chemicals, of which a new build-up had taken place after the war. And the nature of the seed also meant other balances were destroyed, so you needed chemicals also to control pests, and seeds that were hungry for chemicals were also very thirsty for new amounts of water, usually about three or four times as much as the traditional varieties.

Now you needed large irrigation systems through the year, because part of what the green revolution had designed were seeds that could be cropped in multiple ways. They were season insensitive, they were photo-insensitive, so you needed water through the year. You didn't shift from crop to crop, depending on the rainfall and climate patterns. You had to engineer the environment for the seed, and for all this you needed credit, and for that credit you needed to go to the World Bank and other aid agencies. What's usually forgotten when people talk about debt is that before development these were processes of daily living and subsistence in which people didn't have to turn even to their own bureaucracies, forget international development agencies. They didn't have to turn to their cities. They were the producers and there was a clear political base of power in the rural areas based on the knowledge that they were the real producers.

The green revolution gave a new kind of power, based literally on bargaining over prices of inputs and sale of commodities. The only politics it left possible for the farmer was a politics of the market place, of prices. The farmer, the producer, had been turned into a consumer more than anything else. He or she was robbed of a sense of productivity. When the green revolution, because of its inherent non-sustainability, ran into problems and could not maintain yield in the same way, could not maintain profits of the same magnitude, the problem was located not with this process of social and natural transformation but in other political processes.

David Cayley
Is there disillusionment amongst those farmers now? Do they see what you see — the costs of destroying the traditional way of farming?

Vandana Shiva
No, they don't see the costs in that same way. They see the costs much as an American consumer who's getting worried about the environment. They see the costs more in terms of, does this package say it can be recycled? It's the only question a consumer in a supermarket will raise about the environment. Is this package recyclable? The gaze has narrowed down through participating for twenty years in supermarket purchase.

In a very similar way, a green revolution farmer has had his mind narrowed, and by and large the green revolution farmer is a "he," because one thing that the green revolution has systematically done is remove women from agriculture. Punjab is conspicuously different from the rest of India, in that you don't see women on the fields. It has turned women into genuine parasites in that sense of removing them from what is productive activity. Everywhere else you go, you see women doing certain jobs in the fields. In Punjab, you don't see them in the fields.

David Cayley
And you would have, thirty years ago.

Vandana Shiva
Oh yes, you would have seen women working in the fields earlier. The green revolution changed the whole gender presence in agriculture. Now, the green revolution farmer, who is a "he," after twenty years of participating in the green revolution thinks very much like the consumer in the supermarket. He's a consumer in the supermarket of agri-chemicals. A consumer does not ask, Does the forest from which this product comes suffer when it's first harvested? Does the soil in which

these off-season strawberries are grown get hurt? Those are not the questions that are possible because you are not living on those soils. It's not in your imagination. Your imagination, as it's been perked up by environmental consciousness, focuses on the packaging, and your politics then focus on the recycling of packaging and not on the ecological costs of producing for world markets. In a similar way, the Punjab farmer, the green revolution farmer has had his mind narrowed to that interaction between subsidies, prices and production. He does not see ecology as relevant to all this. You go into what I call more backward areas, and there the resistance is of a different quality. They worry about seeds, they worry about returning to a farming that's more under their control in the genetic sense.

David Cayley
Genetic diversity is part of the genius of traditional Indian agriculture. In some areas, nearly 300 different varieties of rice were cultivated, each adapted to some season, soil, pest or perhaps just some ceremonial use. This is part of what is lost when green revolution monocultures take over, and the green revolution is not an isolated case. There is now also an internationally backed "white revolution," so called, in dairying.

Vandana Shiva
The white revolution has the same parallels, that it takes the cow by itself as not linked to the farm. The white revolution takes the cow only for its milk production capacity, and denies its animal energy. A terrible tragedy has happened in south India. Because cattle, just like plants, get bred for maximizing a certain input and a certain output, cows start getting bred for maximizing milk production. Very often, cows that produce more milk are extremely useless for other work functions, and in India, we have always had what are called dual purpose cattle. The cattle need to pull bullock carts, they need to pull plows in the field and they give you milk, and the same breed has to function well for both because the male and female are both useful.

Through the white revolution, the female for the dairy yields is transformed more and more and the male calves become more and more useless. You will get very tasty veal in

Bangalore markets for next to no price because all male cows are butchered, and now there's a total deficiency of bullock power in India, of animal energy, which is creating its own bottleneck, because in paddy growing areas you need to work the fields very heavily to make them right for cultivation.

David Cayley
And the males can't be kept for that purpose because they're the offspring of these specially bred cows?

Vandana Shiva
Of exotics, yes. They're not bred for stamina, they're not bred for pulling weight, they don't have the hump. The Jerseys are flat on the head and our cattle are humped. That hump is what pulls the weight, that's what pulls the plow and the bullock cart. There are many reasons — I am still finding out how many reasons — but many, many reasons. You don't breed male offspring; you basically breed the female offspring.

David Cayley
Can you talk for a minute about the kind of science that makes this possible and your attitude towards the science that has made forestry possible on the high slopes, that has made possible a transformation of the nature of the Indian cow, that has made the green revolution possible, and so on?

Vandana Shiva
To me, it's a kind of a one-eyed science, literally, a one-eyed science that looks only to the market and then tries to design instruments to feed the market. There's a very lovely phrase that one of the women in Chipko used once when I was talking to her. I said, the government is setting up all these research institutions, all these development agencies, and how do you feel about them? And this woman said, "Without food, fodder and water, any thinking, any thought that devotes itself to developing new technologies is a one-eyed thinking."

I call it reductionist thought, in the sense that it cuts itself off from the many purposes that any object in nature or any object that human beings create is supposed to fulfil. It shuts out other dimensions of the nature of being, so it creates extremely efficient artifacts, designs wonderful instruments to maximize flows to the market, and that's the only flow that this one eye can see. But it doesn't have the balancing of seeing what it destroys, let alone being able to serve those other objectives or ends better. It has no capacity. Institutionally it has no capacity. Institutionally it has been trained only to look towards the market. Its very source of nurturance, the nurturance of institutions, their sense of recognition, their sense of relevance, their sense of dynamism comes from how market mechanisms, which push the engine of science, can direct where scientific thought will yield or will not yield.

David Cayley
In your book, you also call this form of thought patriarchal. Today in our conversation, you seem to be emphasizing much more the market as its genesis. Is that a change or not?

Vandana Shiva
No, it's not a change, because various cultures have gone from basically being multi-dimensional to being one-dimensional. It's just like the dual-purpose cattle; the female and the male are needed for different purposes. You don't measure the bullock for yielding milk and you don't measure the cow for animal energy. They have different purposes and they serve those purposes well. From that kind of an existence, we have systematically moved into an existence where men of every society are pulled towards the market, largely through the efforts of other men in centres of power who see the men in the village or the men in the tribal society or the men in the Third World as the head of the household or the breadwinner. So that societies with a division of labour become one-handed societies, and the one hand that links up to the market is the hand of the men of those societies, and the hand that reaches out to these men in Third World societies is a patriarchal hand from centres of capitalist power. In that sense, a science that breeds through the market, given the structure of power between genders, is necessarily a patriarchal science.

David Cayley
It is Vandana Shiva's belief that circumstances are now forcing India towards a more integrated view of the questions of gender, environment and development, which she has been

discussing. Social and environmental disintegration are now so obviously linked, she says, that the very idea of development is beginning to be radically questioned.

Vandana Shiva
It's really in the last one or two years that there's a national level of thinking, organizing and talking about these issues and, in some way, a building of an idea about an India different from where the development push put us. Larger and larger numbers of young people are getting attracted to the environmental movement. For the people whose very survival is at stake when a dam is built or a mine is built or an industry is set up on agricultural land, it's not a matter of choice. It's not a matter of preference. It's a matter of life and death, and for them quite clearly to say "no" to that destruction is the best statement of wanting to live.

So both trends are very strong, and in that sense the genuine environmental conflicts, the real ecological upsurge, if it happens, will take place in countries like India where both sides of the struggle sit next to each other, they live side by side. Part of the reason why Western environmentalism has been hijacked by the technological fix response is because the pressure on nature happens so far away. It's not lived by anyone in your neighbourhood. You don't get the feedback. In India, it's coming all the time. It's coming to the communities that live in the destroyed system, it's coming from the communities into the mainstream each time more people are displaced. And the environmental issue is getting very intimately tied to the issue of displacement, to the fact of uprooting people from their cultures and their locations, and the right to culture, the right to be what you are, the right to live communally is becoming a very major issue of democracy in India. Does a minority sitting in Delhi planning development projects have a right to uproot people in that way? That is one of the biggest and most lively debates.

But what troubles me even more than that is what's happened to society in India with the development process. We were talking about polarized thinking, thinking that necessarily turns the other into a "not-me," and therefore the annihilation of the other is a precondition for the continuation of my existence or my group's existence. Every community, every ethnicity, every region, every religion is going through a kind of identity definition, because part of the development process is the fragmenting of the mind, including the fragmenting of the identity. And that's planting violence on a very large scale in a situation where, just like Illich has said, development is creating its own kind of scarcity. In that situation of growth-generated scarcity and development-generated fragmentation, the combination is extremely volatile, and that's where I personally feel the biggest disturbance in India is taking place right now. The environmental disturbance is very severe but the disturbance to the organic existence of India as a peopled India is even more severe.

David Cayley
Wouldn't you say in the last analysis that they're the same thing?

Vandana Shiva
Yes, they are the same thing. It's just that the environmentalists haven't looked at the other, which is why I state it like that. To me, they're part of the same ecological breakdown.

*** *** ***

Most calls for environmental action are calls for more of something: more science, more environmental assessments, more experts, more money, and so on. The environment remains what economists call an externality, only now, it is said, we must bear the expense of factoring it into our equations. Vandana Shiva belongs to that minority of more conservative thinkers who are calling for less by drawing attention to what is already there. As one example she points to the genius of traditional Indian agriculture, which accomplished by intelligence, adaptation, and attunement what modern systems can only achieve through massive applications of external resources. The modern system may temporarily produce more rice, but it also mines the soil, creates dependency on costly inputs, displaces women, and robs the farmer of his independence, initiative, and inherited knowledge. These problems then require further interventions in support of the broken people and the ravaged land. "An environment" requiring external and expensive therapy has come into existence where before people lived within the bounds of nature's providence without having to attend to this costly "externality."

Vandana Shiva wants to stop this cycle before it begins. Instead of increasing environmental consciousness, she wants to defend a tradition that had no need of this concept. For her, the alienation of the land and the alienation of the people are one. Environmental consciousness is nothing more than salt in the wounds made by development. India, she says, does not need ecology, but a renewed respect for its native genius.

9

Ecology as Design: The Work of John Todd

John Todd calls himself an ecological designer. He believes that the future of civilization lies in living machines, assemblies of organisms that do the work now done by polluting mechanical machines. In a recent article about this concept, he recalls an event from many years ago that planted the idea of living machines in his mind. He and his friend, Bill McLarney, were poking around in the small upland streams of Costa Rica. They came across a species of large fish living in streams that seemed to lack sufficient food to sustain a species so large. They then discovered that these fish were capable of digesting the hard, seemingly inedible fruits that fell into the stream. Closer investigation of their anatomy revealed terrifying-looking teeth, capable of shredding hard materials, and a long, serpentine intestine that was able to digest tough materials. "It dawned on me then," he wrote, "that the world is a vast repository of unappreciated or unknown biological strategies that have immense importance for humans if we could develop a science of integrating the stories embedded in nature into the basic systems that sustain us."

Using ecology as the basis of design has been Todd's life work. Like Shakespeare's banished duke in the forest of Arden, he has found "tongues in trees, books in the running brooks, sermons in stones and good in everything." He began his work in the late 1960s as a founder of the New Alchemy Institute, a pioneer in ecological technologies. Today he directs the Centre for the Protection and Restoration of Waters, dedicated to using ecological knowledge to solve the problems of water pollution and toxic waste disposal.

My programme about John Todd opened on a scene that only a seagull could love — the back end of the dump in the small town of Harwich, Massachusetts. The date was May 1, 1990; and the citizens of Harwich were circulating curiously amongst large translucent cylindrical tanks full of algae, snails, fish and numerous plants. The tanks also contained "septage," the highly concentrated toxic output of septic tanks; but there was no smell, and the building under its gossamer greenhouse roof had a bright, airy, vibrant feel to it. The occasion was the opening of the Harwich Solar Aquatics Septics Treatment Facility, a living machine created by John Todd and his associates, the second such facility they've now completed. Its purpose, as John Todd told the opening day crowd, is to purify septage wastes that are difficult to treat with conventional mechanical and chemical technologies.

John Todd
This particular sludge is extremely concentrated. It's some 30 to 100 times more concentrated than ordinary sewage, so it's very, very difficult to treat. This material also contains, because of our bad household practices, a number of heavy metals and toxic materials, which are in themselves carcinogens. The idea is to purify these compounds without the use of hazardous chemicals in the treatment process. We break up these carcinogens that get into our water, using organisms that have this capability, and try to shunt metals out of the water stream. Inside this building are over 1,000 species of organisms, each of which are working in a constellation to accomplish a task that no single one or small group of organisms could ever do, and that's the reason why it's called ecological engineering.

Ecological engineering is bringing together organisms from the wild and putting them into a new, contained environment to do some of the work for society. In the case here, the work is purifying the wastes. So, in a sense, what ecological engineering and solar aquatics really is, is miniaturizing in a high-light environment the processes that take place naturally in lakes and streams, and doing so under controlled conditions so that we can effect something in a matter of days, say ten days here, that would normally take months in the wild.

David Cayley
The town of Harwich is on Cape Cod, which is essentially a big sand bar extending like a crooked finger off the coast of Massachusetts, south of Boston. There are few sewer systems on the Cape, which means that most wastes are hauled to the town dumps in tank trucks, called "honey wagons," and then dumped into holding ponds or septage lagoons. Below these ponds is the large lens of groundwater that is the Cape's water supply, between them only the Cape's quick-draining, sandy soils. The problem is obvious.

But now Harwich has a solution. It's a result of the town's own political initiative; and this is what is most gratifying to John Todd. He's been engineering elegant ecological solutions to contemporary problems for twenty years. Now local communities are starting to get interested.

The road to the Harwich dump began in the late 1960s, when Todd, his wife and colleague Nancy, and biologist Bill McLarney started the New Alchemy Institute. Medieval alchemy was the precursor of modern science. New Alchemy was to be the harbinger of a new science.

John Todd
I would say in many respects it was quite a sixtyish thing, in the sense that Nancy and Bill McLarney and myself had concerns ranging from the ecological devastation resulting from things like pesticides and abuse of land to our concern with the way science in the broadest sense was going, where so much of its talent, money and energy were going into weaponry. We had a broader sense of inequities — biological, ecological and human — in the world. The New Alchemy Institute was a quixotic but certainly well-intended attempt to see if we could create a science and a practice of earth stewardship. We wanted to redirect scientific activity by applying ecological knowledge to a wide range of human problems. We believed then, and I still do now, that inherent in the world's modern knowledge are the bits and pieces that could create high culture, living in harmony with the planet, and this was

what New Alchemy was set up to do; it was set up to explore these questions of a new kind of science of stewardship and a practice of stewardship.

So we're certainly of our time, but in many ways the questions that we were concerned with in the sixties are the same ones that people should be concerned with today. I don't think that's changed.

David Cayley
In the sixties, the word *ecology* came into general use and became a kind of metaphor, at times almost a manifesto, while remaining at the same time a science with aspirations to scientific rigour, suffering, as Paul Ehrlich once said, from physics envy. Do you see this kind of ambivalence in ecology?

John Todd
I really loved the ambivalence because I was happily in both camps. First, I was trained in straight behavioural ecology with one of the finest ecologists of the time, Marston Bates, and second, I was working with what really were hippie homesteaders who had decided to turn their backs on the crass, materialistic culture of their parents and head off into the wilderness. Only when they got into the wilderness, they discovered they didn't know what to do or how to do it.

So we all became kind of consultants to them, so to speak. But I used to laugh because I would go to scientific meetings and watch the scientific ecologists start to crawl when the ecologists in the social, political, and behavioural sense would arrive in the room. They were very nervous, and very few people could cross the boundaries.

David Cayley
The relations between scientific and philosophical ecologists have fluctuated over the years, but John Todd has always kept his passport and his desire to straddle that boundary. New Alchemy blended science, social vision, and technical ingenuity. It began in Southern California in the late 1960s and was relocated in 1970, to Cape Cod, where it still prospers. There, the Todds and their colleagues began to pose the questions that would eventually produce a whole new family of technologies.

John Todd

One of the first questions we asked was how we could produce the food needed for a small group of people, using a very small space and renewable sources of energy and ecological cycles. In order to accomplish those ends in small spaces, we had to immediately develop integrated systems. And so we had a pond, for example, inside a small solar greenhouse-like structure. The pond was in that structure in order to absorb enough radiant energy so we didn't have to heat it with fossil fuels, much the way the ocean provides the climate for the Earth.

We did the same thing. With that body of water providing the thermal storage and the thermal buffer, we began to use that water column for aquaculture. The work married ancient Chinese methods of polyculture, which we got from the Orient, with modern ecological knowledge, combined with the idea of introducing a lot of light. Out of that grew the whole area of work called the solar aquatic aquaculture. In some cases, on the surface of the water we would be growing foods. To do that, we began to study the ancient Mayan floating agriculture and began to bring that into our thinking.

Then, because we were growing fish in this small space and we couldn't use any agricultural poison — it would kill the fish — we had to find ecological methods of pest control. For example, we had to look at predator-prey relations, a subject dear to the heart of academic ecology, and find those beneficial organisms to fit in our small habitats.

And then we began to start working with wind energy or moving water for providing electricity, for a whole variety of things, and out of that, began designing systems based on pulses. For example, wind doesn't blow all the time, so if you have a fish farm powered by a windmill, you have to design it so that it can cope with natural pulses, which is the opposite to the Western mechanistic point of view, which is to just sock it to it with a continous source of electricity and maintain a steady state system. We decided that working with pulses might be beneficial, and so we started to study pulse-like environments, like tidal marshes, and asked marshes — what are you doing and how do you do it? We wanted to do it too. And that knowledge from mangroves, tidal marshes, and things like that became incorporated into the thinking.

David Cayley
Out of this early research came a new type of building called a bioshelter, a building capable of regulating its own climate, producing food, and recycling its own wastes. Both John and Nancy Todd were born and grew up in Canada, though they've made their lives and careers in the U.S., and they managed to convince the Canadian government to support the building of a bioshelter at Cape Spry, Prince Edward Island.

The Prince Edward Island Ark, as it was known, was inaugurated by Pierre Trudeau, in 1976. As an experimental design, it was a success. As a place to live, it was a little before its time, and the Canadian government eventually withdrew its support. Todd, meanwhile, continued to expand his field of interest, spending time in Costa Rica, where Bill McLarney had established a New Alchemy offshoot, and travelling in Asia.

John Todd
One of the things that was important to me was to go and study parts of the world that had been farmed continuously for millennia, where, you would say, human beings had been doing something right. Now, in the West, farming for centuries is considered a long time. Then the soils erode and the trees are all cut down and people have to abandon it and let it regenerate. The whole history of Europe has several of these phases of internal colonization, and in North America we're seeing a repeat of the pattern. But near Bandung, in Java, in Indonesia, I was able to look carefully at farming systems that had lasted for thousands of years, where fertility had been sustained and possibly increased.

What I learned was fascinating, but the real news was there was an extraordinary balance between water farming and tree farming and grains and vegetables. No one type of farming was allowed to dominate, and in those parts of the world where the ponds provided nutrients and fed the soils, and the trees protected the upper reaches, one can still see a vertically integrated, horizontally integrated, mutually reinforcing type of agriculture.

David Cayley
As John Todd's work progressed, his attention turned increasingly towards the Third World. Faced there with devastated

economies and environments, he began to see applied ecology as the only real alternative to the ruinous type of development that hooks poor countries on manipulated flows of capital and technology. Ecology offered a way of putting people's fates back into their own hands.

John Todd
Only a very small percentage of humanity has access to capital, and the only substitute we have for capital or hardware that capital can acquire is information. It becomes a universal currency. It's the key to creating equity throughout the planet. If we're really interested in helping the Third World, it won't be through the importation of technologies, it will be through some kind of partnership in finding the appropriate information to a given context, and most of that information is going to be biological or ecological.

David Cayley
A case in point occurred in the Seychelles in the early 1980s. There Todd was able to solve a problem threatening an entire island community by applying information originally reported in Russia.

John Todd
A couple of Russians had studied a process in which they had observed that ponds could be found on top of rubble mound hillocks. They said, how could this be? Water should just drain right out of there. What they discovered was a process which occurs in nature which they called "gley" formation. Basically, what happens in nature is if carbon and nitrogen are present in a hollow, and somehow oxygen, over time, gets driven out so it's an anaerobic environment, this biological plastic forms.

So I was in the Seychelles in the middle of the Indian Ocean, visiting a coral island that was threatened. A hundred or so people were going to have to leave the island because their source of fresh water, which was a lens under the island, was being used up and salt water was invading. This process happens on islands in coastal regions all over the world, but it's particularly acute on coral islands because the coral sand has no ability to hold moisture. And so I happened to visit on the brink of a disaster. We asked ourselves if the process observed in those bogs in Russia could work in quick time in the middle of the Indian Ocean?

To find out, we dug a small lake and we looked around for the fibre and carbon that we needed to simulate the Russian process. We didn't know how often it occurred in nature or how long it took. However, we lined this lake with the fibre of coconut husks, which were a by-product of the industry of the islands. And then we went around and looked in the understory of the island for the nitrogen we needed. What we found primarily was a wild type of papaya. We cut the branches, the whole shrub and even the fruit, which also has the active enzymes, and we put in a six-inch layer of papaya. And then the one thing left to do was to drive the oxygen out, and we did that by putting in another six-inch layer of coral sand and tamping it, rolling it down to make sure we drove the oxygen out. Then we took some of the available groundwater to fill it up just enough to cover the bottom. A couple of weeks later, the monsoon rains come, and lo and behold, nature made a biological sealant. And there, in the middle of this island, is a lake. As a result, the people were able to stay there, and it's also acting as an ecological magnet. I understand that Sir Peter Scott visited the island about a year after we did the experiment and there were birds on their way from the steppes of Asia to Africa that were landing there that never land on coral islands, somehow attracted to the source of surface fresh water.

So that's an example of knowledge derived from studying wasted places and finding ecological processes that counterbalance the natural tendency for a place to become wasted. That knowledge, which was used to serve a human community, made a fundamental difference.

David Cayley
When I first interviewed John Todd in 1981, he described for me a vision of what he called "a biological hope ship." The ship would carry the biological materials needed to restore shattered ecologies or even, as in the Seychelles, create altogether new ecologies. She was to be called the Margaret Mead, after the late anthropologist who had been Todd's friend and mentor, and was to be a great sailing ship, inaugurating a new age of sail. He wanted to combine his passions for farming and

sailing, he said, and sail a farm around the world. It was a grand vision, but Todd had trouble finding investors for his sailing symbols of the age of ecology. Eventually, he settled for something more modest — a sail-powered fishing boat. By this time, the Todds had left the New Alchemy Institute and started a new organization called Ocean Arks International. Their first project was stimulated by the plight of Third World coastal fishermen.

John Todd
We were aware, both through direct observation and through FAO United Nations reports, that literally millions of fishermen in the last few years have been unable to fish, that they no longer have the fuel for their outboards or the capital for their gear. This is a world-wide phenomenon, that coastal communities are in really tough shape. And if you analyze why they're in tough shape, it's because their countries' economies have lost their buying power. Their currencies have become worthless. There are very few hard currencies in the world. We are used to being able to take a dollar and go anywhere with it, but these countries cannot. For example, one country with which I've had direct experience is Guyana, South America. A Guyanese dollar won't buy you anything, and what happens when these currencies go soft is that the infrastructure disappears. Mr. Yamaha outboard motor pulls out. The local distributor doesn't have a hard currency, so he can't buy it. The oil importers don't want to import oil because no one's going to pay for it with anything.

So what happens is the economies come unglued. We saw this happening everywhere, at a terrifying rate, and so one immediately says, Well, they could go back to the old ways. They could build traditional boats and do it the way they did it a generation earlier. That argument breaks down when you discover that the old ways involve boats being made of teaks and mahogany and rot-resistant woods that have all been cut down to pay for the outboard motors and the steel boats and everything else. The trees are no longer there; their biological capital has been used up.

Consequently, I started to ask the question, Would it be possible to build a boat that could be powered by the wind, that could be built primarily out of fast-growing, soil-restoring weed trees, that would be as fast as the motorboats they were intended to replace, that would have the most advanced aerodynamic and hydrodynamic thinking? Could we take the information from a high-performance aircraft or speedboat, or what have you, and apply that to the needs of artisanal fishermen?

David Cayley
The answer to all of these questions turned out to be yes. With the help of a prominent naval architect named Dick Newick, Ocean Arks came up with a vessel which they called an "ocean pick-up," a one-and-a-half ton trimaran. A prototype, called the Edith Muma, was built in Dick Newick's boatyard on Cape Cod and launched with great fanfare in November, 1982. Dean James Morton of the Cathedral of St. John the Divine in New York gave a revised version of the traditional Anglican prayer for the launching of ships, the champagne was gently poured, not broken, over the hull, the Paul Winter Consort played "Amazing Grace," and shortly afterwards the three-hulled Edith Muma set sail for Guyana. The Guyanese fishermen were sceptical but eventually impressed.

John Todd
A funny thing happened in Guyana. We had a little radio on board that nobody knew about. And we would sail into Georgetown, and we'd hear a trawler captain say, "Mon, that three-winged thing is worthless," and someone respond, "Yeah, mon, I'd like to see you take that tub of yours all the way to New York," and they'd argue back and forth about the merits of it and we'd sit there, listening on their frequency. Then occasionally we would sail in and the trawlers would try and catch us, but with the Trades, we could go faster for sustainable periods of time, and we would listen to them laughing at each other as they chased us into Georgetown, telling each other to put a little bit more power to it. Thank God the shrimpers were a national company because they were burning fuel like crazy to stay up with us. It was satisfying.

David Cayley
The fishermen who sailed with Todd and his crew also liked the ocean pick-up's performance. One offered to buy her the first day out. The boat was designed to be manufactured on a

small scale, out of mainly local materials, but it still required some local investment and that was where things broke down.

John Todd
There was a lot of interest in Guyana in building a large fleet of several hundred of these, because the one thing Guyana does have is marine resources. The financial community was interested in a technology transfer and it looked promising. But there were certain sectors in the government that saw this as liberation technology, if you will, and they didn't like the idea of a thousand fishermen being able to go anywhere. They didn't like the idea that they could sail up to Trinidad and pick up spare parts for the remaining outboards that were still functioning, or slip down to Surinam to get wheat flour and bring it back, because wheat flour is illegal. They were perfectly happy to control the movements of that sector of the society at the gas pump, and I was struck by the sort of shadow side of the internal combustion engine which is the ability to control people, and that's how they seemed to do it. The dominant sector in the government, which is, as you know, not a democracy, decided that this is not what they wanted.

David Cayley
The Edith Muma set sail for Costa Rica, but her problems winning acceptance continued. "Small is not beautiful," Todd has written about this experience, "when it comes to research vessels achieving recognition by the appropriate authorities." Ocean Arks joined forces with the Marine Research Centre at the University of Costa Rica, but co-operation proved difficult. The ocean pick-up remains an idea whose time hasn't yet come. There is still no prospect of the boat being manufactured, nor any governmental interest in Todd's larger vision of the ocean pick-up as part of an integrated scheme of coastal development. This experience changed his orientation. He still believes that ecological technology is critical for poor countries, but he's concluded that it will have to prove itself in the rich countries first.

John Todd
When I was working in the Third World, people asked me, if this is such a good idea why are you not doing it at home? At

that point I said, they're telling me something I have to pay attention to. I will go back and do solar aquatics, develop fleets for New England, or vessels that would be appropriate to New England, develop architecture that could be used in Canada or United States, and give myself ten years doing that, get that understood as the way of the future, so that one could go to Jakarta and say, "My God, Massachusetts already has twenty-two of those and 200 more in the works," Then I will have the credibility. Too often it's said, what are you bringing, second-class knowledge to the Third World? We don't see you doing it at home. And then you get labelled as a do-gooder rather than a person of creative action.

David Cayley
One of the first projects that Todd got off the ground was a solar aquatics waste treatment facility for the city of Providence, Rhode Island, not far from his own home on Cape Cod. Like the Harwich facility described earlier, it's a greenhouse structure, containing engineered marshes and lines of translucent tanks where an amazing variety of plants, microbes and marine creatures purify Providence's wastes. Unlike conventional secondary treatment plants, it uses little energy, no hazardous chemicals and produces no toxic by-products. It was opened in July, 1989, and has continued in successful operation since. I asked John Todd to take me for a tour.

John Todd
We're right in the bowels of Providence, Rhode Island, in the centre of the most industrial district of the city. To our immediate east is 63 million gallons of partially treated sewage roaring out into Narragansett Bay. To our immediate north is a dog pound where all the stray animals of the state gather for their last days, and off to the west is one of the largest haulers of liquid asphalt and other nice materials in the state. To our immediate south, overlooking Narragansett Bay, we have a glorious mountain of scrap metal, which is destined to go back to Europe, where they turn it into BMWs.

And right in the middle of all this, we have a gossamer-like greenhouse structure, and inside that structure is a water garden. If you were to walk into it, you would see ginger and flowers and watercress and fish and snails and clams and herbs and spices. What comes in at one end of the building is raw, untreated industrial sewage from an industrial city, and what leaves the building is pure water that has been transformed from that original state.

Inside the first chamber, raw material sewage is quickly being transformed into great vats of algae, and then the algae, in turn, are kept in check by large grazing populations of snails, which you can see here. All these dots over the surface are snails, which are the sheep and the cattle of the aquatic realm. The organic material is burst up to the surface, the surface is grabbed on to by the roots of these microscopic floating plants here, and in there is where the bacteria reside that do the waste treatment. Then the snails consume the bacteria that treat the waste, and so begins the basis of the food chain.

These mountains of foam that you see coming off are the various surfactants and wastes from restaurants and households, soaps and various types of things, and they foam up, and in those foams, again, there are algae and bacteria and other organisms that also continue the treatment process. At the very beginning, you can see higher plants floating on the surface. That's a tropical plant, the water hyacinth, and during the summer months here, it is just a mass of orchid-like flowers. The contradiction between the treating of waste and the aesthetic is one that we find very interesting and somewhat ironic.

We've now moved into the second and largest room inside the solar aquatic waste treatment greenhouse. Basically, this second stage in the treatment process mimics the strategies a salt marsh uses, which is a period of drying and a period of wetting. For half of the day in here, the marshes dry out, and that allows air to penetrate down into the system, and then the other half of the marsh — they're all in parallel, there are eight of them here — the other half of the system is wet and it becomes anaerobic and does one type of purification process. The side over here, which is dry, becomes aerobic and is prepared to do another kind.

And as you can see, this is an eclectic marsh, if there ever was one. We have the umbrella plant, which I think originated in North Africa; we have the eucalyptus, which is from Australia; and we have three or four species of the scirpus or bulrushes, from North America. These are all intertwined in this system to produce a polyculture, with each plant having different depths and different functions. Some remove organic carcinogens, actually physically break them up. Some of these other plants entrap heavy metals and lock them up rather than allow them into the environment, and get them out of the water so that they're not re-released into the bay.

What we do with these heavy metals varies, depending on the plants that they get in. If it's a tree, a long-lived tree, then we like to find trees that concentrate metals in the stems and roots, and we can lock up these metals then for centuries or many, many years because they're planted out after they start life in this building to become landscape trees afterwards, and that's an important side of the story. Other plants take up heavy metals, like the tiny floating plants, and those are composted before they reach hazardous levels in the plant, so that they can be re-used on the landscape, but the metal levels are low enough that they don't cause long range degradation of the external environment. The third strategy that we use here is unique, and that is because this is an industrial city and the backbone of this city is in fact the jewellery industry. What we're attempting to do here is try and find plants which are called hyperaccumulators, and these are plants that actually try and mine specific species of metals out of the water and concentrate them so that they could be re-used as ore grade. And we have a long-term project to try and study just this phenomenon and see if we can find metal mining plants that are happy to live in water or in wet marsh-like environment.

David Cayley
Does the sewage you treat have significant amounts of precious metals in it?

John Todd
Yes, all of the precious metals are here. We don't measure for gold, but we measure for silver and then there's cadmium and mercury and lead and other metals like those. It's all in the sewers of the city and it's very erratic. Sometimes there's very little and then all of a sudden there'll be a big spike, and that tells us that one of the factories has done a discharge into the sewers of the town.

As we walk toward the final treatment process, the actual diversity of plants and animals increases. We come into an area where we have crayfish and clams and different species of plants. It's a tiny aquatic fern floating here that produces that carpet-like mat and then, of course, there's the ubiquitous watercress in this system, which are the work horses in here.

The final stage, which you see right here, is basically again an engineered marsh, but it's really a polishing marsh. The idea is to remove the last of the fine particles, and the other thing that's very important in this phase of the process is to have plants in the polishing marshes which are powerfully antibiotic. And most bulbs are antibiotics, and that's one of the reasons that they store well and don't rot easily if they're kept relatively dry, and so you'll see a fair number of bulb plants in the system. Irises tend to be very good this way. Then the water passes down through the fine crushed gravel filter and then leaves the building as clean water, roughly four days after it entered here.

David Cayley
After our tour, John Todd and I continued our conversation in the quietest place we could find — the cab of his pick-up truck — rolling down the windows after each exchange to prevent heat prostration and then rolling them up again to keep out the noise of the passing trucks. As we sat, facing the greenhouse, Todd told me what they had learned from it so far.

John Todd
The facility we're looking at here can treat the needs of about 150 households. It's roughly eleven by forty metres in length. If we were to do all of the city of Providence, we would need an area of approximately 120 acres. That's comparable to the acreage that is currently used by the city to treat their waste to secondary standards, and this facility is treating it to advanced wastewater standards, so when it comes to space, these new, light-driven, ecologically based processes are space competitive.

Consequently, the opportunity to treat the whole city in one place is here. But the other side of the coin, which I think is very important, is that because it's beautiful, because it doesn't stink, we now have the opportunity for the first time to disaggregate the problem of waste. So that each neighbourhood or each community could have its own facility, each neighbourhood or community could use its own by-products, the trees, the various flowers and things like that, to enhance the environment, so that these facilities could become epicentres for the landscaping of whole areas, including cities. The state of Rhode Island, which is our host for this particular facility, is very definitely interested in the idea of disaggregating their problem, and the reason is, if anything were to go wrong in the city's current plant, we would have somewhere between 60 and 100 million gallons of raw sewage a day going into the Narragansett Bay, one of the great bodies of water in this part of the world, whereas if one of our facilities was serving a community there would never be that kind of disaster. So that's the value of this type of ecological engineering, that it can help disaggregate the problem because it's no longer a foul process that nobody likes to think about.

David Cayley
What about the cost of doing it?

John Todd
We've addressed the issue of cost in two ways. The first is if we're dealing with a very concentrated waste, like septage, which is 30 to 100 times more concentrated than sewage and hard to treat conventionally, then we are far more cost effective than any other technology. We're way out front there. When it comes to sewage, we don't yet know our costs in relation to other facilities. It looks to us as though, for the price of an ordinary secondary treatment plant, we can build an advanced wastewater treatment facility. It looks to us as though we are modestly more cost effective in the dilute waste or sewage waste area.

David Cayley
A final question about this plant. Knowledge that you've spent half a lifetime acquiring must be at play here. Is this a coming to fruition of the knowledge that goes right back to the beginning of the New Alchemy Institute?

John Todd

There is no question that this is a fruition, a coming of age. I have enough experience with different kinds of organisms and different plants and different animals and the way they work together in concert, that now I've actually reached the point in my life where I can start talking about something quite revolutionary and quite new, and that is the concept of a living machine.

A living machine is an ecologically engineered technology that uses wide varieties of organisms to carry out the work of society. I can see the same kind of knowledge being used to produce foods without any environmental degradation, and perhaps environmental enhancement. I can see the same kind of living machines to produce fuels for our automobiles, to regulate our climates, both heating and cooling and air purification. So in areas of waste, food, fuel, and even architecture, one can begin to see how living machines, which are contained in these gossamer-like environments, with light penetrating everywhere, can function as work horses. In a sense, for the first time in the history of technology we're able to actually miniaturize the process of production and recycling. And some of these living machines can be made to last for centuries, unlike mechanical or chemical engines. Some of the simple parts might wear out, but the overall systems can go on forever. They're self-replicating, self-repairing. They have all of the capabilities of machines, except they have attributes that machines don't have, hence the name "living machines."

And so I think we're on the threshold of something really fundamental, and carried one step further, in fact, we are even beginning to talk about major projects, both in eastern Europe and in New York City, of actually designing intelligent buildings with all of their support functions carried out by living machines. We're stepping into a new dimension, which is interesting, because it's fourteen years ago that the Ark on Prince Edward Island was finished, and that was the first complete statement of an integrated system. It's interesting that what we were doing was totally misunderstood for about fourteen years, and now the pieces have come together.

David Cayley

Ten years ago, in a book called *Tomorrow Is Our Permanent Address*, John and Nancy Todd drew a distinction between the structure of a system and its coefficients. An automobile is a structure, and the fuel efficiency of its engine a coefficient. Tinkering with coefficients is the easiest and least threatening way to approach environmental problems. Improving the fuel efficiency of cars without challenging the structure of transportation is a good example. Using energy to recycle something that needn't have been produced in the first place would be another.

John Todd's eye has always been for structure. He wants to redesign society so its structure resembles the structure of the living world. In living systems, each part is linked to the whole, but retains a certain independence. This is what gives the system its resilience and adaptability. Our bodies are a co-ordinated play of such relatively independent parts. In contemporary social structures, that element of autonomy is missing. Analyze even your household economy and you'll probably find yourself linked into scores of unstable, ecologically destructive, politically questionable supply lines extending right round the world. Living machines are John Todd's answer, a way of making civilization continuous with nature by designing as nature designs.

One of John Todd's inspirations in this work has been the Gaia hypothesis, the idea that the earth, as a whole, is self-regulating. Gaia, he thinks, is the framework in which ecological engineering finally makes sense.

John Todd
The notion of the earth as alive is ancient, but when the idea becomes part of the consciousness of people, then their sense of the earth changes dramatically. And my sense is that economies built on ecology will allow people to live and believe in one system, whereas now you can believe in Gaia and a single, wonderful manifest ecology, but how do you act on that belief, how do you live on that belief? And I think sometime in the future, the living and believing can come into harmony, and if I didn't think that, then I would probably have very little hope because I'm aware of the damaged ecological fabric of the planet. I guess the idea of Gaia is coming around at the right time to provide a broad mantle under which people change the values and the way they work and their relationship with

other living things, not just with each other. So that's my source of optimism.

*** *** ***

John Todd stands at the opposite end of the spectrum from Vandana Shiva in the previous chapter. Shiva defends traditional forms of subsistence; Todd invents entirely new forms, although he often draws on traditional knowledge. His innovations are extraordinarily cosmopolitan, combining organisms from all over the earth, using knowledge gleaned from Russian bogs to make new lakes in the Seychelles, and blending high tech with tradition. He doesn't deprecate tradition, but he points out that, where people's "biological capital" has been used up, tradition can no longer guide them. Restorative action must involve the engineering of new ecologies, and the marriage of traditional forms with critical bits of high technology.

In a sense, Todd stands outside the development versus tradition debate. He is radically modern in his desire to remake the world, traditional in his insistence on designing as nature designs. His sewage treatment plants are like nothing ever seen before, and yet they could be deployed on a village or neighbourhood scale. His vision is planetary, his action very precisely local. He offers hope that, despite its vexing contradictions, the age of ecology may be a habitable place.

10

Gaia: A Way of Knowing

This programme profiled scientist James Lovelock and cultural historian William Irwin Thompson. Just as Thomas Huxley was once called "Darwin's bulldog," Thompson has been called Lovelock's bulldog for the way in which he has popularized and explored the theory which Lovelock and microbiologist Lynn Margulis co-authored: the Gaia hypothesis, which takes its name from the ancient Greek goddess of the earth, Gaia, the mother of the gods. The name was suggested by the novelist William Golding. Golding and James Lovelock were neighbours at the time that Lovelock conceived the idea that planet Earth might constitute a single cybernetic system, and when Lovelock explained the theory to Golding during a walk round their rural English village, Golding proposed the name. It was a portentous choice. The evocation of Mother Earth and the ancient religion of the goddess gave the theory a cultural resonance lacking in a bald scientific statement of the idea that the earth is self-regulating. The idea of Gaia dovetailed with feminism's recovery of the Goddess, inspired artists like musician Paul Winter to create his Missa Gaia, or Earth Mass, and helped focus the concern of environmental movements on the planet as a whole.

The theory, in a sense, was overwhelmed by its own cultural implications, but Gaia remains primarily a scientific hypothesis which holds that life on Earth produces and regulates its own environment, or better, that life on Earth is its own environment. The origins of the Gaia hypothesis lie in some work Jim Lovelock did for the U.S. government's National Aeronautics and Space Administration in the 1960s. It was during the planning phase of NASA's Viking mission to Mars, and Lovelock and his colleague Dian Hitchcock were asked to devise experiments that could detect the presence of life on Mars, should such exist. Lovelock, with the naïveté of genius, decided to turn the question round and perform a thought experiment designed to detect life on Earth.

James Lovelock
We thought, well we better check our theory by looking at a planet that does have life on it, and of course the only one we know about is the earth, and it's quite easy to do a *gedanken* experiment and set up an infrared telescope on Mount Olympus and look back at the earth. So we did this, and when we looked back, we found an atmosphere that was wildly anomalous, and a strange, wonderful and beautiful anomaly that sort of shouted a song of life, as I said, right across the solar system, right out into the galaxy.

If somebody says, What do you mean by this, what anomalies? I say, well just consider two of the gases, oxygen and methane. Oxygen's present at 21 per cent, methane's present at 1 1/2 parts per million, a mere trace you say, but their coexistence at a steady state in an atmosphere represents an anomaly measured in hundreds of orders of magnitude, as far as its disequilibrium goes. For methane and oxygen to coexist in an atmosphere on a planet at that steady state means that something must be making the methane and something must be making the oxygen, because they react together and they use each other up. Knowing the volume of the earth's atmosphere, you can calculate that the something must be introducing no less than a thousand million tons of methane every year into the atmosphere, and something must also be introducing something like 4,000 million tons of oxygen every year into the atmosphere to account for the losses from the reaction of these two substances. And there just aren't any non-living processes that can do that in an atmosphere such as the earth's, so the answer must be that there's life.

So we reported this to our sponsors, NASA. They couldn't have been more disgusted. You see, not only had we proven that there wasn't any life on Mars, and they badly needed life on Mars to justify sending Viking there, but much worse than this, we'd used NASA funds to prove that there was life on Earth, and they were scared witless that the message would get back to Senator Proxmire, and you can just imagine the questions he would ask about this waste of NASA money.

Of course, he would have been wrong. It wasn't a waste of money, because looking at the earth that way was as much a scientific revelation as the view that astronauts have. When the astronauts first saw the earth, many of them said, "My God, the thing must be alive, it's so beautiful, it's so much a whole."

But what we were seeing was a hard science suggestion that there must be life. You see, to keep those unstable gases at a perfect steady state requires a lot of organization, but much more remarkable than this, how on earth could an atmosphere that was a bit like the gases that go into the intake manifold of an internal combustion engine be just right for life? This was even more extraordinary, and of course that's what made me think, well maybe we're looking at it the wrong way round. The atmosphere isn't an environment for life, it's something that life has made as an environment for itself. It's something it has chosen and deliberately keeps going because it likes it that way. And that, of course, was the Gaia hypothesis and that's how it started.

David Cayley
At the time that you were having these thoughts, what was the mainstream scientific thinking about the origin of the atmosphere?

James Lovelock
Much as it is now, that it was pure geology, that the biota was just a passenger on the planet and had very little to do with it. We just used the oxygen, and we burn the carbonaceous matter and we return CO_2, and the plants take in the CO_2 and push back oxygen. It just goes round and round in a cycle and does nothing, said the geologists. Life has no effect on the geological evolution of the planet. They're so locked into their

paradigm that they don't seem to be able to realize how inconsistent their position is.

The question I always ask is, What would happen if all life suddenly ceased on earth? What do you suppose the atmosphere would be? And they rarely ever give a straightforward answer, but you can quite simply calculate it and model it, and you find that in the course of perhaps a million years — it takes a long time for geological process to go through — we would finish up with an atmosphere very like that of Mars or Venus. It would be dominated by CO_2, there would be very little oxygen at all, probably no nitrogen, certainly no methane, and the planet would probably be very hot indeed. Not as hot as Venus, but getting far too hot for life.

David Cayley
Can you explain some of the Gaian mechanisms — perhaps the oxygen-methane cycle?

James Lovelock
I could, but that's a more difficult one. Let me explain one of the ones that we know best about, and that's the CO_2 one, because there's a lot of contemporary interest in that, too. You see, one of the more convincing bits of evidence for Gaia was the constancy of the climate throughout geological time. For 3-1/2 thousand million years — the time that life has been on Earth and nearly a third of the age of the universe — the temperature has been constant, the climate's been constant, and yet the sun has been steadily warming up, and this is one of the strongest arguments in favour of regulation. So how did it happen?

One geochemist, Jim Walker, tried to explain it on purely geological grounds. He said, or rather accepted, the geological evidence that right back in the beginning, when life started, there was a great deal of carbon dioxide in the atmosphere, perhaps as much as 30 per cent of the atmosphere was CO_2, and that's what kept the earth warm and enabled life to get its start. He then said that the simple process of weathering — that's the reaction of carbon dioxide with calcium silicate rock, which removes carbon dioxide from the air and deposits it in the sea as limestone — would account for a steady diminution of CO_2 over time, which would exactly equal the rate of rise of solar luminosity. It was a nice theory and a good try, but when you put the numbers in it, it wouldn't work. And I thought that he'd done exactly the right thing. The only thing he'd done wrong was to leave life out.

You see, life is very much in the business of weathering, of rock digesting and so on, and Jim Walker's process can be made to work beautifully if you put life there. If you analyze the soil in most places on the earth, you'll find that its carbon dioxide content is thirty times higher than that of the atmosphere. So on the soil, in the soil everywhere, life is pumping CO_2 out of the air as hard as it can in order to get it to react better with calcium silicate rock and get Jim Walker's reluctant chemistry to proceed. In other words, Gaia facilitates the process that the geologists had envisaged, and without life it wouldn't happen. And in the sea, the same process is going on. The sea is continuously pumping CO_2 out of the air and down to the bottom, a sort of conveyer belt, and without their pumping, CO_2 would rapidly rise in concentration and make the earth uninhabitable by living things. Earth is a feedback system that operated all the way back to the beginning.

David Cayley
The objections that have been made to your hypothesis by Ford Doolittle of Dalhousie, for example, centre on the fact that it offends conventional Darwinian notions of how natural selection operates, because he claims that there would have to have been foresight or planning amongst the biota which is denied *a priori* by the theory. How do you respond to that?

James Lovelock
His criticism was understandable in the context of biology as it interprets Darwin nowadays. It's just like the geologists I spoke of earlier. They live in a paradigm that does not see a world where the environment and life are so tightly coupled as to constitute a single cybernetic system. They see the evolution of the species as taking place almost independent of the environment. The species may adapt to changes in the physical environment, but they don't see that the evolution of the different species automatically changes the environment, changes the rules of the game in which the next species is going

to evolve, and that this tight coupling is what makes Gaia work.

David Cayley
Jim Lovelock first formulated the Gaia hypothesis in the 1960s. One of the few scientific colleagues who took his idea seriously was American microbiologist Lynn Margulis, and they have collaborated ever since. She filled in many of the details of the theory from her studies in microbial evolution, describing how microbes have altered both the atmosphere and the surface of the earth and emphasizing how much more important symbiosis and co-operation have been in evolution than competition.

Ten years later, in the 1970s, this work caught the eye of cultural historian William Irwin Thompson. As a cultural historian, Thompson believes that science is always embedded in some larger story, and he saw in the Gaia hypothesis the scientific narrative that could knit together a planetary culture.

Bill Thompson is director of the Lindisfarne Association, a loose affiliation of thinkers devoted to fostering a planetary culture. The association is named after a Celtic monastery established in the seventh century, off the coast of Northumbria. Lindisfarne was one of the centres from which the medieval Christian civilization spread through Europe, a place where the ideals of classical civilization were "miniaturized," as Thompson put it, into a curriculum for a new era. The new Lindisfarne, founded in 1973, was to be a seed of planetary civilization. It became the intellectual vanguard of the New Age movement at a time when "New Age" still meant something more than crystals, channelling, and feeling good about yourself. He is also a poet, essayist and author. His thirteen books range from *Evil and World Order*, published in 1971, to *Imaginary Landscapes: Making Worlds in Myth and Science*, published in 1989. In the same year, Thompson brought out a collection of essays called *Gaia: A Way of Knowing: Political Implications of the New Biology*. In this book, he states that "Ecology will be the political science of the future," and Gaia the sign of a new way of knowing the world.

William Irwin Thompson
The quintessential idea in Lovelock is that worlds embrace repulsions, and that processes that seem to be violently op-

posed can be constitutive of other architectures of order, so one animal's excrement becomes the food for another bacteria, and the planet is a delicately balanced thing between the fixed and the fluid. The continental plates are fixed, more or less, over time. They are also another kind of fluid, but in a short-term perspective, they're fixed and the gaseous atmosphere is fluid.

Consequently, a healthy living system like us, with our fixed skeleton and our fluid rivers of blood, has to embrace these opposites. If we don't, then we come up with defensive mechanisms of trying to crystallize value into a gene, a subatomic particle, a museum, a currency, a metal, and all of these ideas of value fixed in objects are perishing everywhere you look in the culture, whether you look in art or whatever. And Lovelock and Margulis are a good way to get a handle on that idea and understand it. For me, they're the quintessential shift from ideological thinking to ecological thinking.

David Cayley
But what does the Gaia hypothesis say to our more traditional ideas of nature, the objective existence of nature?

William Irwin Thompson
First, it says that nature is an arbitrary threshold. You cut a square in the universe and you stand on the bottom of the square and you call that threshold the window, and on the other side nature. But where are you going to cut that square? Are you going to do it at the molecular level and see the entrancing dance of molecules and flashing electric skins and light that might be at another threshold pollution? This beautiful vision you're having of the dance of molecules in nature might be a New Jersey toxic dump, but if you're inside it, at the molecular level, it could be wonderfully natural. Or you could be at a level of a supernova exploding and just creating havoc, and that can be nature, too.

When we say "nature," then we're really influenced by Sierra Club calendars, and the photographs of Elliot Porter and Ansel Adams, which are influenced in turn by Constable and Gainsborough. Our concept of nature is derived from an eighteenth-century gentlemanly vision of the great estate and the park, and it's been given to us by great city planners like

Olmsted in creating Central Park in New York City. It's a cultural idea. It has nothing to do with nature. In the nineteenth century, nature was objective and the observer was subjective, and had no value. All value therefore came by decreasing subjective contamination to achieve a reading of nature that was pure and true, and the most pure was where human interpretation was least present, as in reading a meter.

We now have the same thing, only we call it deep ecology. We think that nature is at its purest when it's not contaminated by trailer parks, not contaminated by weekend hikers, not contaminated by selling pharmaceuticals from the Amazon rain forest for Ciba-Geigy companies, or something of this sort. But there is no such thing as that nature; that's a fiction. Nature is the horizon of culture. Every time you change cultures, you change the horizon. So nature in a shamanic culture might have angels and elementals and spirits. Nature in a cybernetic, cyberpunk landscape might have machines that were ensouled by entelechies.

David Cayley
"Cyberpunk" is a literary genre, working the blurred boundary between reality and its simulations. The best known example is William Gibson's novel, *Neuromancer*. Gibson's characters have electrodes implanted in their skulls. They jack in and cruise the video landscapes and virtual realities of what Gibson calls *cyberspace*. *Entelechy*, in this context, means a soul or guiding spirit.

Willliam Irwin Thompson
Imagine, for example, the cyberpunk worlds of *Neuromancer*. Here we're going into the science fiction landscape of the "unnatural." In a Hopi culture, you would take the molecular lattice of a sacred mountain and a holy spirit would ensoul the holy mountain, and then the shaman going into meditation would commune with the mountain and have a vision. In a cyberpunk landscape, the molecular lattice of a cybernetic organism would be ensouled by an entelechy and the Druid wizard who was jacking into cyberspace would begin to commune with the spirit that ensouled that mechanism.

For us, in our nineteenth-century romantic world, we think nature is trees and mountains but not that other technology, that is abhorrent, that is evil and that is unnatural. But I think if one really wants to understand what's going on in the shifting horizons of our culture, one has to understand nature as going in two directions simultaneously. One is the return to nature, with the Greens, and the other is the destruction of nature in the cyberpunk landscape of things like *Blade Runner* or *Neuromancer*. And unless you look at both of those edges of our culture and ask yourself, What is nature, I don't think you'll really come up with the transformation that's going on right under our nose.

David Cayley
The obvious difference between the Hopi shaman's mountain and the cyberpunk landscape is that the one is human made, the other is not.

William Irwin Thompson
Well, we didn't make the silicon and we didn't make the electrical pulses and we didn't make the laws of physics and nature and entropy and all those other things. Let me give you Lynn Margulis's example of what nature is. She said all the environmentalists come to Boston and they look at Boston Harbor and say it's dead. It's polluted, it's unnatural. And she says, But, I see all my "friends" out there — meaning all the bacteria that she studies — and they're chewing the tires and they're frolicking in the oil slick. So I have a sense that it is an arrogant consumer's nineteenth-century aristocratic image of nature that we're talking about. Nature is a fiction. The only precise way to define nature is by saying that no such thing as nature exists. Nature is the horizon of culture, and whatever you are in, in whatever human activity, you will always have a horizon.

David Cayley
Then, how can ecology provide the moral dimension in political science, which you said it would?

William Irwin Thompson
Because ecology is actually studying processes within our horizon. I didn't say there wasn't a horizon to our consciousness. Ecology studies how a cell works, how a swamp works,

how biological processes enter into a dialogue and interact with human beings. The biologist, Rene Dubos, would say, if you look at the horizon of Florence, you see a man-made artifact. You are seeing the beautiful Tuscan hills and the vineyards, which have been sculpted by humans in the same way that a beaver would create a dam, or in the same way that a bacterial mat would create a stromatolite, which is kind of an artifact left from bacterial activity. The Gunflint Schist in Ontario is actually the remnant, the iron ore is the remnants of the oxidization processes of bacteria from zillions of years ago.

Consequently, with Gaia the division between animal, vegetable, and mineral breaks down. There is not a wall between them, but a shifting and highly permeable membrane. So you don't want to say the mineral is unnatural. Do you want to say nature only begins to be nature when there are animals and trees, or do you want to take it back to the origins of life? But what about before life evolved, the pre-biotic soup? What about the mineral period, the Hadean epoch, before we had even pre-biotic molecules kicking around in the ocean? That has to be seen as nature. But if we study these things we can see that it's an industrial era mentality to come in and level Kansas and put in wheat, when the prairie operates in a different way and has a more complex dialogue. Nor should we romanticize it and say the prairie was pure when the Indians were there, because they had prairie fires, and they burned out a lot of the higher vegetation and they, as much as we can tell, would create stampedes with fire to have all the animals fall off a cliff and have a huge slaughter. It's called the extinction of the Pleistocene mega-fauna. My point is that there were intrusions on nature in 9,000 or 8,000 B.C., when Gary Snyder's Indians were the only inhabitants of this continent. They were changing nature and they were sculpting the prairies. And so, everywhere we look in "nature," we see processes like that going on. They then teach us about how culture works and how we understand the interrelationship of opposites in a condition of health.

David Cayley
The Gaia hypothesis shows all creatures actively constructing their environments and, in effect, becoming environments for each other. A tree doesn't just stand in the environment, it

creates the environment by entraining the forces of wind and water, developing the soils, and making a home for a myriad of other beings who, in turn, serve the tree. This Gaian perspective offers an understanding of evolution quite different from the classical Darwinian view.

Thompson's close associate, Francisco Varela, suggests replacing the term "natural selection" with the words "natural drift" in order to eliminate the fiction of a stable environment that can do the selecting. In his essay, in "Gaia: A Way of Knowing," Varela illustrates the concept with a translation of a poem by Antonio Machado. "Wanderer," the poem says, "the road is your footsteps, nothing else. In wandering, you lay down a path. Turn back, and the path there is none, only tracks on ocean foam."

William Irwin Thompson
Let's go back to this old way of thinking. The object in the container, the organism in the niche. Now, that's a way of thinking in biology that none of my colleagues would accept. What they now see is that animals, through their metabolic processes that are shared in a common phase space, extrude the evolutionary landscape so that their excretions and their inhalations, create a kind of dialogue through time, and so they're climbing on top of one another's niches. And one will create a form of pollution that's a disaster, and the other one will scurry around very quickly and then slowly begin to adapt to, say, the presence of the oxygen excreted by the cyanobacteria, and then that begins to change the atmosphere, and then organisms begin to change with the atmosphere.

So the dance of life is now seen more in terms of what Varela calls natural drift rather than adaptation. The old notion is that you have to adapt or you're going to die, so identity is in a gene that you can manipulate and there's an organism that must adapt to its niche, which is clamped into its niche. In the other Prigogine kind of biology, the organisms are actually dancing and they are extruding their environment, so it's like a river that is changing the banks at the same time that the banks are sculpting the river and the river is sculpting the bank. You have to change your language, so Varela uses lovely poetic language like "brought forth," "worlds are brought forth," or you use concepts like "emergence." So the particular evolutionary landscape that is brought forth is radically different at each particular time, and nature is changing throughout this process.

David Cayley
Since you've mentioned Varela, and he's been an important colleague of yours, can you tell us who he is and what the new biology you've spoken about is?

William Irwin Thompson
Varela's voyage is an interesting one in terms of planetary culture. He started out, as a teenager, reading Heidegger in German in Santiago, as a kid who grew up in a mountain village in the Andes. He got his PhD at Harvard at age twenty-three and then moved in the seventies into studying Tibetan Buddhism with Chogyam Trungpa Rimpoche in Boulder and came to live at Lindisfarne as a scientist, scholar in residence in the 1970s. And we've been working together on a programme for biology, cognition, and ethics over the last three years and have written four books together that are in the process of appearing. So Varela is definitely a colleague and he's been one of those who's greatly influenced my thinking, and he enabled me to try to make the connection between cognitive science and the Gaia hypothesis and helped me to build a bridge between ecology and biology and cognitive science and political science.

Let me give you an example of what I mean. In the old days, cities were charismatic, and we thought that élites would be in the city and particular cities would carry the civilizational energy for a time: think of T. S. Eliot in London or Jean-Paul Sartre in Paris. And now, right at the time that we're beginning to think in terms of Gaian processes at large, of how the circulation of the plankton in the sea affects the formation of the clouds, affects the albedo, and how the reflection of the solar radiation affects the temperature of the planet, we're beginning to see that civilization is no longer *civitas*, it's no longer located in the city. It's a distributive lattice, which is a concept that comes out of cognitive science, and Varela is a cognitive scientist. He's a neurophysiologist. And this is a particular term that comes out of a branch of cognitive science that's called connectionism. It is devoted to trying to create slower computers with fuzzy logic that think analogically, rather than fast computers that work digitally with just gates of off-on, one-zero, and the only way they can achieve this complexity is through parallel distributive processing. And so the ideas in the brain are not simply located in one cell, they're a distributive lattice that organizes the whole brain into a domain or a state. Parallel distributive processing and connectionist lattices and emergent states that have the capacity to learn are precisely what we're talking about when we're dealing with Gaia. Gaia is a system of learning that maintains itself over time.

Varela has also studied the immune system, and from another point of view, you could define Gaia almost as the immune system of the planet that maintains itself and its self-identity over time. So that if you look at Gaia at the atmospheric level, with Lovelock's work in atmospheric chemistry, you look at the macrocosm. With Lynn Margulis, you look at the microcosm of the bacteria. Margulis will argue that bacteria are not distinct species, they are one super-organism, a planetary bioplasm, which is an idea that's been developed by Sorin Sonea in his new bacteriology in Montreal. That gives you a planetary bioplasm and if you study the immune system in the individual, that gives you a particular entity that isn't a discrete object but is like an enclouded self that is maintaining through the blood and through the marrow a definition of selfhood over time, where the self really begins to be the phase space of the body in the same way that if you study the movement of the Foucault pendulum, its phase space is larger than the ball. And so the concept of dynamics and learning and how a metadynamic can emerge from a highly connected system so that it begins to be self-naming, autonomous and maintain that autonomy and identity over time begins to be really fascinating. In order to understand these processes, because they are processes, and not objects, you have to say I need a new geometry to be able to perceive these because my old geometry always asked me to look for an object. But this is saying no, an object is not a phase space, that you need to think not in terms of Euclidean geometry but in terms of chaos dynamics.

David Cayley
"Chaos dynamics" refers to a new science able to make mathematical models of complex forms: a wave breaking, smoke

rising from a cigarette in turbulent air, the flow of air over the wing of a bird in flight. It belongs to a new phase of mathematics able to describe the interacting processes which comprise Gaia. This heady brew is what Thompson called "the new biology." I myself find it somewhat unnerving because it pictures a world without a ground, without, as Varela says explicitly, "a privileged perspective." It offers no way of clearly demarcating the human from the continuum of life. Inside Gaia's magic bubble, where identity derives from a process and not an object, all boundaries seem permeable and impermanent. Out of all this Thompson conjures the vision of a new politics, based not on turf and egotistical interest but on what he calls "noetic ecologies," temporary structures of shared information like, say, the Live Aid concerts for African famine relief which dissolve, disappear and reform like clouds. I wondered aloud during our conversation about the sense of home in such a world.

Do you see any danger of losing rootedness, embodiment, sense of place by adopting this ecology of consciousness?

William Irwin Thompson
Well, I guess that's why I've always been involved in contemplative practice because I had this argument with Wendell Berry once at Lindisfarne in Colorado. Wendell is a close friend as well as a Lindisfarne fellow, and we've all been thinking out loud in these jam sessions for the last twelve years. And Wendell was going on about this rootedness and the spirit of place, and his family have been in Henry County, Kentucky, for nine generations.

And I'm more of an electron than a nucleus. I don't have a location, so I am almost Wendell's exact opposite, and so I have a tendency to feel that I am deracinated, that I am unnatural, that I am unrooted, that I have no sense of identity, that I'm your typical uprooted academic nomadic intellectual. So I remember feeling frustrated and I said, "Damn it, Wendell, you keep talking about place, but I see the monarch butterflies heading for Mexico and I see the humming-birds leaving me for the winter, and over the horizon I imagine the whales heading south, you know. It wasn't the rich that invented this lifestyle, it was, quote, animals in nature. So what is all this stuff?" And I said, "As a matter of fact, the nineteenth-century farm is perhaps a much more disastrous imposition on nature, that if we go back to 9,000 B.C. before agriculture had gotten fixed with surpluses, we have a seasonal round in gathering, and we have hunting and gathering and fishing, and we don't have values so much fixed in location." It was the increasing surplus of grains that allowed us to start holding food in containers and then surrounding our buildings with walls, and then men could take their hunting bows and arrows and use them for raids, and raids grew into warfare.

So you can say that agriculture, the fixing of value in turf, is inseparable from militarism, and that to say that identity is only valuable in fixed values means you're going to have to get your AK-47, it's me and the missus and my rifle and everybody else is a threat. So I don't draw my identity from fixed turf or from my meat body. I find my identity much more involved in very complex topological processes that move in more than three dimensions, so this enables me to live in a way that might be disorienting for someone who's on a nineteenth-century family farm.

David Cayley
The debate between Bill Thompson and Wendell Berry is one that now divides the entire environmental movement. As Donald Worster and Wolfgang Sachs have both argued, ecology has always been an ambivalent field of thought, containing both a modern and an anti-modern mood. As science, it's modern. As a romantic reaction to science, it's anti-modern. Thompson has committed himself unequivocally to science and has faced the consequences: the end of any stable or permanent idea of nature, including human nature. His choice highlights the much more conservative mood of other elements in the environmental movement. Thompson calls the Greens, for example, a nativistic movement which sees the past as the future.

William Irwin Thompson
Part of their project, with Rudolf Bahro and others, is to go back to a pre-industrial society and try to recover archaic terms. They don't want to deal with the world of chaos dynamics and mathematics and big science and space probes to

Mars, or whatever. They have a very reactionary view. It's basically a nativistic movement.

David Cayley
It seems to me that at one time you yourself may have held these contradictions together, too.

William Irwin Thompson
I'm sure I still hold contradictions because you couldn't have a brain or a complex personality unless you embraced opposites. The creative process is inherently a dance of opposites, it's a complex kind of alchemy, and any simplifying ideology always falsifies one side of our nature. So I did set up Lindisfarne as a younger person more naïvely, as a nativistic movement. It would be the revival of the humanities in an age of technology, and since I couldn't do it at MIT, I quit MIT and came to York University in Toronto, and then found the vision for the development of a modern university at York was Stanford and MIT, to basically boost the Ontario economy, and that was not what I had in mind.

So then I quit and set up Lindisfarne and it became captured by too much nativism, trying to go back rather than forward. It was not the prophetic imagination, it was the regressive one, and the whole New Age movement is full of that. Neolithic matrilineal agricultural villages, palmistry, dowsing, Zen, Tibetan Buddhism, these are all cultures where we've been before. This isn't new stuff, this is old stuff. I mean, it's good stuff but it's old. It's called the "New Age," but it's not new.

And so, it was at that point that I began to be aware that in the dynamic of understanding the complexity of change, imagination was too limited to what Marshall McLuhan would call looking in the rear-view mirror. To really get a sense of the horizon, you had to ask yourself, What am I afraid of? What terrifies the hell out of me? Where do I see evil and the unnatural? This is why I became fascinated with cyberpunk landscapes. One begins to see where the emergent change is really occurring. And if one has a prophetic imagination, then one begins to realize that the transformation of evil into good, the Christic transformation of the Mephisthophelean that Goethe studied, has always been with us and is most likely to occur in those areas where we're afraid. And once we walk through our fears and understand the disintegration of nature, the end of nature, understand that we're at the edge of the flesh, then the novel and unexpected becomes possible.

David Cayley
Thompson's realization that his own Lindisfarne Association had been, in part, a nativistic movement was tied to what he was seeing in the new biology: a world in which time and change produce constant novelties, a world without an external Archimedean point from which it can be viewed, a world where the sacred means something very different than in the traditional schools Lindisfarne was originally intended to revive.

William Irwin Thompson
I began to understand why the school that I had actually helped bring together and found funding for, the school of sacred architecture, made me unhappy. I remember an argument with Keith Critchlow in London — who is not a reactionary, he's a kind of Summerhill-educated, English Labour party socialist, but he's very committed to the Platonic idea. And he pointed to his watch and said, "The centre is fixed," meaning all the world of temporality and change and appearance just goes round and round. And so, there was the idea that nothing is ever new and that values are fixed.

I had a kind of deep experience in meditation where all of that just died for me. It was like a real death experience, and I could feel the melanoia where my mentality changed and I moved out of that way of thinking. Intellectually, I began to understand and appreciate that there was a new mathematics on the horizon that was part and parcel of the true New Age and was not this medieval Platonism. I had grown up with Whitehead as my high school culture hero, and I was always a Platonist. At that point, in 1983, I diverged and split and resigned from the New Age and began moving. I began to be much more interested in my association with people like Ralph Abraham and Francisco Varela and less in the earlier project where I had been working vigorously with Keith Critchlow and Kathleen Raine, very much with the idea of returning to the past in a kind of Yeatsean romanticism.

David Cayley
The original Lindisfarne was a monastery, a cultural enclave in an era that history called the Dark Ages. And during the late 1970s, in books like *Darkness and Scattered Light*, Thompson did see the new Lindisfarne in essentially monastic terms. He now rejects this view.

William Irwin Thompson
When I set up Lindisfarne to preserve the humanities in an age of darkness — the idea was, we have to preserve knowledge in an age of change or loss — what I didn't appreciate was that the metaphor was right but the content of little enclaves off in Auroville, Findhorn, or us out at the end of Long Island was misplaced concreteness and was too literal. We were going through a period of cultural loss with incredible degradation of literate culture: I saw that we would end up with "McBooks" and the New York editorial élite becoming just a commodity marketing thing, and *Time* magazine decaying to the level of *People*, and newspapers coming down to the level of *USA Today*, and television programmes like *Sesame Street* ensuring that children didn't develop an attention span. So we've gone through an incredible period of the loss of literate culture. Now, in that sense, Lindisfarne was an attempt to bring together artists and scientists and poets and painters and folk to hold on to some levels of culture in a period when we were just getting the Shirley MacLaine of everything, and so its model was defensive of identity. It was the profane again. It's us versus them, and I think that was not imaginative and inappropriate on my part. It was too narrow and didn't understand the larger process.

David Cayley
Bill Thompson's thought has always been based on the insight that historical outcomes are inevitably paradoxical. Human beings, by definition, can never know what they are doing because the rational mind can only illuminate one thing by obscuring another. The world is therefore "a structure of unconscious relations," as Thompson says, and planetary culture can only be the result of a process apparently driven by terror and greed. The Gaia hypothesis offers Thompson a physical cosmology in which these paradoxes of history make sense.

Gaia is the larger systemic mind of which we are the unconscious parts. We cannot be conscious of this greater mind by definition, but we can identify ourselves with it, and it is this identification that animates Bill Thompson.

William Irwin Thompson
If one adopts a longer historical view, one tends to lose one's sense of value and location. How can I be motivated to go out tomorrow and join the Sierra Club or Greenpeace or whatever? The larger scale of time is disorienting if one founds one's identity on an ego with an agenda. In point of fact, I'm not on that pH scale at all. In order to be empowered to act and to do what I feel is of value, I don't think it has to occur in my own lifetime. I'm perfectly willing to involve myself in a project where I may never see the results of the activity. That was the original reason for calling my association Lindisfarne, because the monks of Lindisfarne, in 635, didn't live to see the cathedral of Chartres. If you're dealing with a shift from one world system to another, that's not in the clock time of an individual life. So I don't involve myself in either pessimism or optimism. I find that a more contemplative sense of the big picture is actually empowering. If I took too narrow a point of view, I'd really get bummed out because I'd be only looking at the short term thing, and the short term thing always shows you bad guys 99, good guys zero. But if one looks at a larger picture and says that once we were eukaryotic bacteria and once we were dinosaurs, and then we were hominids and then we adopted this contradiction between animals and apes that we like to call human, and now the human is ending and we're moving into some end of nature and end of human nature with it, and it's beyond our imagination but it involves a revisioning of indentity and value and politics and science and everything else, I find that empowering rather than disempowering.

When writing came in, it was a threat to oral culture and it was seen as a threat to memory. There is a quote of Plato's where he talks about writing as the attack on memory. And we go through a period of darkness, and then after a couple of centuries we get, lo and behold, something called sacred texts, and we suddenly get Upanishads and the Bible becoming a Torah and a canon, and now the sacred is invested in what before was evil.

There's a wonderful story that a friend of mine, a physicist at Lindisfarne named Lou Balamuth used to tell, of two people standing by a stream, watching ants on a log flowing down a turbulent stream, and the ants keep moving their position to stay out of the water. And one guy says to the other, "Gosh, look at those ants move," and the other says, "Yeah, and those ants think they're driving that thing."

*** *** ***

A good deal of reservation has been expressed in these pages so far about "planetary consciousness" or "global thinking." Wolfgang Sachs, for example, has expressed concerns about the erosion of culture, the substitution of "ecocracy" for politics, and the supercession of communities of citizens affiliated in a vision of the good by a planetary anthill of "populations" bent on survival. William Irwin Thompson gives a more positive account of planetary culture, not so much by denying what Sachs or Wendell Berry see, as by insisting that this is only the shadow of a larger process which must be contemplated whole. The Gaia Hypothesis, with its suggestion that catastrophe is always at the same time a new evolutionary opportunity, offers him the necessary long view.

Thompson suggests that we are in transition from an age of ideology to an age in which ecology will form the basis of political science. Societies based on turf, ideology, and objective bodily existence will give way to a culture without boundaries, where people identify with processes not objects, change not permanence, and the interpenetration of good and evil rather than a simplistic conception of some unalloyed good. Both nature and human nature will be recognized as purely relative concepts. Thompson offers, in effect, a mystical complement to the banalities of those who would manage the planet: the planet which manages itself without regard for the ants who think they're driving it. And since, for Thompson, we clearly aren't driving it and Gaia is only "a state of learning that maintains itself over time," he sees no alternative to reconciliation with a process in which we ourselves represent nothing more than a transitory phase.

11

David Ehrenfeld: Stewardship and the Sabbath

In the late 1940s, shortly before his death, the American conservationist, Aldo Leopold, published an essay called "The Land Ethic." In this essay, he raised disturbing questions about the utilitarian, human-centred approach to conservation in which he himself had participated as the author of an influential text on game management. "One basic weakness in a conservation system based wholly on economic motives," Leopold wrote, "is that most members of the land community have no economic value. When one of these is threatened, and if we happen to love it, we invent subterfuges to give it economic importance." To get out of this bind, Leopold proposed that society be centred on something greater than the human interest, what he called "the land community," of which humanity was to be no more than a "plain citizen."

Leopold's essay raised questions which are more pertinent than ever today, in the midst of widespread panic about the environment. Is the earth ours to manage? Do humans actually have the capacity to manage it, in any event? Is an environmental movement which adopts the utilitarian language of economics trying to drive out the devil with Beelzebub?

One of the books which raised these questions for contemporary environmentalists was David Ehrenfeld's *The Arrogance of Humanism*, first published in 1978. Ehrenfeld is a professor of biology at Rutgers University in New Jersey, the editor of a journal called *Conservation Biology* and a well-known writer.

David Ehrenfeld
What originally made me write *The Arrogance of Humanism* was a paper that I had written called, "The Conservation of Non-Resources." In that, I examined the problem of what do we do about the 90 or 95 per cent of animals and plants in the world that don't have any value to human beings that's obvious.

Do we pretend that they have a value, do we concoct values, do we search and see if we can find values or do we develop other reasons for conserving things that don't seem to have any value and in fact never had one?

David Cayley
David, what approximately did you and do you mean by *humanism*?

David Ehrenfeld
The way I use the word *humanism* in the book, it's one of these "motherhood" words. It has so many meanings and some of them are things that you can't possibly argue against or dislike. The definition I used is about the second or third definition you'd find in the dictionary, which is making a religion or the religion of humanity. It's the belief that human control knows no bounds, no limits, that ultimately we are the be-all and end-all on this planet and we should therefore have faith in our own abilities to arrange things as we see fit. That's the humanism that I was referring to.

David Cayley
Can you give an example or examples of what you mean?

David Ehrenfeld
An old example is the Aswan dam. The Aswan dam was built to solve a particular problem, which is that they needed power for industrialization, and of course there were political problems too because the Russians were building it for the Egyptians and there were political reasons why it had to be built. But they were told before it was built that it was going to cause all kinds of health problems because the irrigation canals would

have snails which would spread schistosomiasis — which is a terrible disease — all over Egypt. A number of Harvard medical school parasitologists were essentially kicked out of Egypt when they warned about it.

In addition, the dam stopped the flooding of the Nile basin so all the spreading of nutrients brought down by the river over the soil, which was a free spreading of nutrients during the flood season, stopped. The dam is silting up, as dams always do, so that all of that nutrient which was at one point useful when the river was spreading it itself is now just junk, sitting at the bottom of the reservoir and making the reservoir shallow. It cut off the flow of fresh water to the eastern end of the Mediterranean, which made the Mediterranean more salty, and yet it reduced the nutrients at the same time, and made the algal growth less common, so the sardine fishery died. And one can go on and on. The dam was an unmitigated disaster for Egypt.

David Cayley

The Aswan dam is a classic case of unwanted side effects, foreseeable to some extent, but ignored in the pursuit of the main chance and eventually overwhelming the intended benefits. There are dams just like it all over the world, dams with silted up reservoirs, dams whose turbines are choked with water hyacinths, dams which drove whole peoples from their homelands and broke their spirits. The history of foreign aid is full of such projects. But Ehrenfeld's point is not restricted just to mega-projects like dams. He thinks the models of biologists are just as likely to go awry as the models of engineers. One example, and it's one eastern Canadian fishing communities are likely to be sensitive to at the moment, is the concept of maximum sustainable yield in fisheries biology.

David Ehrenfeld

If you start fishing in a fishery, at the beginning, at least when the fishery is first fished, you actually can get more out of it as you fish more, and there are probably a variety of reasons but one of the reasons, for example, is that you're catching the older fish which are hogging a lot of the resources but not growing very fast and therefore not leaving resources — food — for the younger fish which are growing quite quickly. So you can actually increase the yield of fish caught just by fishing a fishery, up to a certain point, and that point, theoretically at least, is the maximum sustained yield. You can also, in theory, continue fishing at that level forever and always catch that level of fish.

This is the theory. It was nicely exploded about ten years ago by a Canadian fisheries biologist by the name of Philip Larkin. What Larkin did in a paper called "An Epitaph for the Concept of Maximum Sustained Yield," was point out that the idea treats a fishery, a species of fish, let's say mackerel or herring, as the only thing in the sea. But of course there are many species of fish and other kinds of animals and plants upon which the fish ultimately depend, all of which are interacting, and this complexity makes it impossible to deal with a fishery as if it were composed of just one species.

So in fact, when you manage one species, another one that's valuable may go down, or things that are happening with the second fishery may affect your plans for the first one. It really kind of gets out of hand. In my book I pointed out that this was very reminiscent of something that John von Neumann and Oskar Morgenstern, a great nuclear physicist and mathematician, and economist respectively, had pointed out in their book on the economic theory of games: namely, that in a closed system, you can't maximize more than one variable at a time. It's just not possible to do. There are limits, and this is one of the limits. I think we should be suspicious, whenever we hear that our activities in the environment are working out just fine when they involve a great deal of control, because very often they don't.

David Cayley
David Ehrenfeld's fundamental point in *The Arrogance of Humanism* is that "the best laid schemes o' mice an' men gang aft a-gley." His approach in this sense seems to resemble Ivan Illich's. Illich has identified a phenomenon he calls "paradoxical counterproductivity," whereby institutions, once they cross a certain threshold of size and intensity, begin to frustrate and subvert the very purposes for which they were established in the first place. Education stupefies, medicine sickens, the machine turns on its creator. Ehrenfeld sees similar inherent limits to successful human intervention in the environment, and feel-

ing this way, he's sceptical of the current rah-rah, "we can turn it around" approach to environmental clean-up, feeling that it may not have grasped just how deep the problem goes.

David Ehrenfeld
I don't think there's any doubt that if we do not change our fundamental philosophy and our approach to dealing with this world, that all the recycling, all the clean-up, all the neighbourhood committees, all the river watches, all of this sort of thing in the world will not be enough to make even a dent in the problem. It really will be just a tiny blip on the history of environmental collapse. That sounds very bad. If these remedial kinds of actions, clean-up actions, are accompanied by what I would call some spiritual action, then I think we have a reasonable chance. But without it, I just don't see any hope at all.

David Cayley
Setting aside, just for the moment, the spiritual action necessary, why will these efforts be only a "blip"?

David Ehrenfeld
Because if we say it's going to be life as usual, except that we will try to clean up our little piles of dirt as we go along, the problem is just hopeless. The environmental crisis is of much greater magnitude than that.

There has been this rah-rah spirit in conservation, and it has been applied to the saving of species. It's important to try to save species in zoos, and some of the more responsible zoos like the Bronx Zoo, and the Lincoln Park Zoo in Chicago, and the San Diego Zoo are certainly doing that. Nevertheless, it's quite clear that we can't save more than a trivial percentage of animals in zoos and if we do save them in zoos, what have we got? What is a tiger that has been kept in zoos for three or four or six generations? What kind of animal is it? Is it still a tiger? Is it a large pussy cat? Does it know what to do, genetically, in the wild? Is it capable of coping with Siberian winters or Indian monsoons? We don't know.

We're trying to save seeds of endangered plant varieties in places like the National Seed Storage Laboratory in Fort Collins and in places like Kew Gardens in England, and it's a failure. It's an abysmal, stinking failure. For a whole number of reasons, we cannot save seeds of even the varieties of things that we have created in this world. And, in fact, we often are losing more than we're acquiring, so every time a new variety comes in, on the average, an old one disappears, of corn or wheat or rice or eggplant or whatever we're trying to save. But there are biological reasons, as well as the political and technological ones, why this kind of saving doesn't work. What we have to do is protect the farmers who are growing things in the environments in which they live. In other words, we're really talking about a kind of problem that technology is utterly incapable of coping with. It's too big for technology and too complicated for technology. We just don't know what to do, how to do it, nor do we have the resources even if we did.

So I would say the spirit of Earth Day is wonderful, provided we have a mechanism for translating it into the realization that, as Wendell Berry says, we have to all learn to live a little bit poorer. We have to learn to live without ruining, and that is going to mean that there are things we cannot do any more that we seem to want to do.

David Cayley
Living poorer, for Ehrenfeld, means living on an entirely different scale. Like many ecologists, he sees that environmental destruction has proceeded at all times and in all kinds of social systems. Ancient civilizations wrecked their agriculture, just as modern civilizations are doing. Communism, as we now see from the sick children and sterilized soils of eastern Europe, is worse than capitalism. Ehrenfeld concludes that the large-scale state is itself the problem, however it is organized.

David Ehrenfeld
The social system, at least in the classic sense of socialism versus capitalism, does not make a lot of difference. What does make a difference is the degree to which a society decides it's going to be managerial. If you set up a large-scale centralized management, regardless of the political system, whether it's a democracy or a dictatorship, whether it's pure socialism, pure communism, pure capitalism or some kind of mix, you're going to have the same kind of environmental degradation. On the other hand, if you set up a system in which your

political units and your control units are small, fairly decentralized, and somewhat hands off, you're going to have much less environmental degradation than you do now.

So I think that there's going to be a great shift which we're now seeing the beginnings of. The paradigm that we've all been brought up with is communism versus capitalism, but that stuff is old hat. It's not productive and it's not useful. The next paradigm that's important is big versus small, centralized versus decentralized, control versus hands off. This is the paradigm that the next century is going to have to cope with somehow. How, I'm not sure.

David Cayley
David Ehrenfeld's denunciation of human arrogance, like his call for spiritual action, has deep roots in the Jewish tradition from which he comes. He denies the prevalent view that the biblical religions are the source of human chauvinism towards nature. This view traces back to an influential essay written by historian Lynn White Jr., in 1967, called "The Historical Roots of Our Ecological Crisis." White argued in this essay that Christianity in particular had preached man's destiny to dominate and exploit nature. David Ehrenfeld disagrees.

David Ehrenfeld
Yes, there are the famous two sentences, two verses in Genesis I, verses 26 and 28, in which Adam is told to go out and take dominion over the earth and to subdue it. And that, according to Lynn White, gave a license to Christians to go and destroy. In fact, however, those verses were never interpreted, in that way, either by the early Jewish sages or by the Christian church fathers.

Let me read to you an extract from Ecclesiastes Rabbah, which is a commentary on the Book of Ecclesiastes, first redacted in the eighth century. Now, this commentary was written down 1200 years ago and it is probably older than that. At any rate, the point is that this period was not a time when people were worried about the environmental crisis.

> In the hour when the Holy One, blessed be he, created the first man, he took him and let him pass before all of the trees of Garden of Eden, and said to him, 'See my

works, how fine and excellent they are. Now, all that I am going to create for you I have already created. Think about this and do not corrupt and desolate my world. For, if you corrupt it, there will be no one to set it right after you.

Think of the power and grandeur of this. But these people were writing in the eighth century, in what we call the Dark Ages. How in accord is that with the thesis of Lynn White that the early Jews and Christians and modern Jews and Christians have taken a license to destroy from the Bible?

Here's another little commentary. This is from the Talmud, the great Jewish commentary on the law, and it is just four lines:

> Our masters taught man was created on the eve of the Sabbath, and for what reason? So that in case his heart grew proud, one might say to him, "Even the gnat was in creation before you were there."

Isn't that an extraordinary statement? In *The Arrogance of Humanism*, I very carefully considered this article of Lynn White's and so I used two quotations, one to start the book and one to end. And the quotation that I started with was from the Book of Job: "Is it by your wisdom that the hawk soars and spreads his wings toward the south? Is it at your command that the eagle mounts up and makes his nest on high?" Here God is saying to Job, I created this, you didn't. Who do you think you are? And then I ended the book with a brief quotation from Isaiah, and this is a modern Jewish translation, and I think a good translation of the Hebrew: "It was your skill and your science that led you astray and you thought to yourself, I am, and there is none but me." That sums up my point, what I'm talking about when I say that we have to recapture some kind of spiritual dimension in our relationship to the world, and a little bit of humility, too.

David Cayley
This raises a question about where our attention should be directed. There's a lot of discussion about saving the planet, which seems to me to direct attention outward. And I wonder

if that's good, whether we can deal with this without directing attention inward, without seeing that it's we who are being corrupted and not just the environment as a sort of a colourless, tasteless, odourless "out there."

David Ehrenfeld
Yes. I'm sitting here with a book at my elbow by Wendell Berry, *The Unsettling of America*. For many of us, Wendell Berry is the first and last word on the subject of where the world is heading and where it ought to be heading. And Berry has always said that conservation begins at home, that environmentalism begins at home, and this idea is absolutely critical. One has to put one's internal house in order, and then go to the community, and if there's any luxury of time or energy left over, then you go on to wider things. Some people have to have in a sense some of that time and energy left over because there has to be some spreading of this idea around the world and some communication. But first you start at home and then it has to extend from oneself. You can't be a hermit and be an environmentalist, just as, for instance, you can't be a hermit and be a practising Jew. You have to have a community.

David Cayley
Environmentalism is a movement which seems divided in many ways but which ranges certainly from a managerial perspective — an attitude that is confident that sustainable development is possible — to a biocentric perspective, which is unwilling to put the interests of human beings above the interests of other beings. It seems to me that coming out of your Jewish roots, you take a different view, neither one nor the other.

David Ehrenfeld
Let me try to answer your question by describing the Jewish attitude towards work and the Sabbath, which, for me at least, is the ultimate way of stating this problem. In Judaism, you're supposed to work six days and rest on the seventh. On the seventh day, on the Sabbath, you are supposed to stop working and there are three things you have to do if you are going to observe the Sabbath correctly. You can't create anything. I

mean anything. If you get an idea for a book, you cannot write it down on a piece of paper. That's very painful for an author and it happens to me all the time, and I wonder, will I remember this till after sundown on Saturday, and sometimes I do and sometimes I don't, and I have stopped worrying about it. If you're a gardener, you can't plant a seed. That's a creative act. You can't do it. You also can't destroy anything. Again, if you're a gardener and you see a weed growing in your garden, you can't pull it up, you can't kill an insect pest, you can't shoot a rabbit, or anything of that sort. The third thing that you're supposed to do is a positive injunction, which is to celebrate the Sabbath and celebrate the fullness of the earth that was given to people to live in, to work in, to enjoy. So you have this prohibition against creating or destroying, which means you cannot be a manager, you can't be a steward, even in any sense. You've got to leave nature alone, and it will continue all by itself. It's a wonderful lesson. You also have to learn how to enjoy it, and that's the other part of the lesson. People were told to have the confidence in the earth and in the creator of the earth that says I'm going to just rest for one day, I'm going to leave it alone.

I think that stewardship without the idea of the Sabbath is bound to go wrong. Without the idea of the Sabbath, without some idea of a built-in restraint, then the steward eventually becomes very arrogant. Hence my title, *The Arrogance of Humanism*. The steward says, I'm really the king. The late J. R. R. Tolkien, in *The Lord of the Rings*, set up this dilemma of a steward who says, How long do I have to stay a steward if the king doesn't show up? When do I become king? And the man who asks this question is told by his father, who is the steward, Even ten thousand years wouldn't be enough, and essentially there is never a time when a steward becomes a king. Well, I think that there's a great temptation for stewards to want to play king, to want to play God, and we need some kind of a restraint that's built in at a regular basis, a kind of constant reminder you're not running the show, you can't run the show. You don't know enough to run the show and you never will and you're only going to mess it up if you have that attitude. Without that idea, stewardship is bound to go awry, to go amiss. I think that the idea of the Sabbath, for Jews, and perhaps for Christians too, introduces this idea of restraint which is so essential to keep stewardship on the right track. So I think that stewardship is the only hope, but I think it has to have some kind of restraint built into it.

*** *** ***

David Ehrenfeld's discussion of stewardship points to a path which I mentioned in my introduction to this book: the way of humility. He begins from the religious assumption that creation is given to us to enjoy, not improve; but he also offers practical reasons why our attempts to improve it may very often run afoul of Murphy's Law: if things can go wrong, they will. He sees part of the answer in a change of scale so that we don't generate problems which are "too big for technology and too complicated for technology." But the final answer must lie in our making some equivalent of the Sabbath: a way of seeing that the world can be safely left alone.

V
Redefining Development

12

A Mental Ruins

Economics and ecology are inseparable, and the question of development is woven through earlier chapters in "Citizens at the Summit" and "From Commons to Catastrophe." Here, it is dealt with explicitly. *Development* is a term that comes originally out of biology, where it refers to the unfolding of organisms, or in evolutionary biology, species. Only in the twentieth century does it come to signify a universal pattern of economic growth, obeying transcultural laws and transferrable from one society to another. The idea of non-Western societies as "underdeveloped" does not appear until after the Second World War; and it is only with the concept of underdevelopment as a remediable condition that the contemporary crusade for international development is really born. It was announced by U.S. President Truman in his inaugural speech in January, 1949:

> We must embark on a bold, new program for making the benefits of our scientific advances and industrial progress available for the improvement and growth of underdeveloped areas. More than half of the people in the world are living in conditions approaching misery. Their food is inadequate. They are victims of disease. Their economic life is primitive and stagnant. Their poverty is

a handicap and a threat, both to them and to more prosperous areas. For the first time in history, humanity possesses the knowledge and skill to relieve the suffering of these people. The United States is pre-eminent among the nations in the development of industrial and scientific techniques. I believe that we should make available to peace-loving people the benefits of our store of technical knowledge, in order to help them realize their aspirations for a better life. And in co-operation with other nations, we should foster capital investment in areas needing development. Such new economic development must be devised and controlled to the benefit of the peoples of the areas in which they are established. The old imperialism, exploitation for foreign profit, has no place in our plans.

In 1949, international development must have had a bright innocent sound, at least for Truman's American listeners. The U.S. had emerged from the war as the world's pre-eminent power. It needed a way to exercise its leadership without the taint of colonialism. What better way than as "Partners in Development," as the Pearson Commission on International Development was later called.

Today, as development staggers into its fifth decade, the very meaning of the word has become uncertain. Development implies a destination, an image of what a developed society is. Truman did not doubt that the U.S. was the proper object of the world's desire. Today, with Western societies facing growing poverty, aimlessness, and incivility, they can no longer be objects of unambiguous admiration. The obvious damage that development has done to the natural and human fabric of these societies has made clear that their way of life is unsustainable. This has created a crisis for those who want to renew and invigorate the project of international development, so they have invented "sustainable development" to provide the new conceptual framework they require. But there are others who claim that development itself is an obsolete idea. It's too compromised, they say, too weighed down with contradictory meanings to be of any further use.

The first chapter of "Redefining Development" contrasts these two views by juxtaposing conversations with David Brooks of the International Development Research Centre in

Ottawa and Wolfgang Sachs, whom I introduced earlier in the "Age of Ecology" section of the book. Sachs, who spoke first, was by this time a research fellow at the Institute of Advanced Studies in Essen, Germany. In the early 1980s, he edited a journal called *Development*, published in Rome.

Wolfgang Sachs
I served as the editor of *Development* for approximately four years, not because I felt myself to be a Third World expert, but rather because I wanted to represent the historical experience of the West with progress, with modernization, to Third World representatives, and I wanted to pull these experiences into the general discourse on development — what development is all about. Today, you cannot discuss development without taking into consideration that those nations who had long been thought to be the ideals of development, have ended up in a dead end. The moment we talk about development, we necessarily include the image of a fully developed society, and historically this has been the United States and Europe. Now, the moment we no longer know what a fully developed society is, there is no point in talking about underdevelopment.

So the whole conceptual framework seems to be crumbling today. It is time to admit that we have here in front of us what I would call a mental ruin. It's time to look at this mental ruin, to examine the layers upon which it is built, to be amazed at the kind of structures, the kind of buildings, the kind of annexes, and the shape, in order to leave it behind, to say, well, this has happened in the past. This we can say in order to explain to ourselves what has happened, but now it's time to leave the shadow of these ruins. So in order to be better able to say farewell to development, I thought it would be useful to have an archeology of it.

David Cayley
From when do you date the concept of development?

Wolfgang Sachs
I would like to modify the question slightly. I would rather say where I see the concept of underdevelopment beginning. And that is surprisingly clear. If you look into the *Oxford English Dictionary*, you will see that it was President Truman, in his

inauguration speech on January 20, 1949, who used first the term "underdeveloped areas" in this world. Before 1949 this was not known. One did not speak in that term about countries, let's say, in the southern hemisphere.

David Cayley
But surely the colonial powers had always regarded these same areas now being called underdeveloped from a fairly lofty height. Did they not always regard these areas as underdeveloped?

Wolfgang Sachs
No, I don't think so. Of course, the colonial powers looked down upon these countries. However, it was a looking-down-upon which somehow comes out of a different attitude. It comes out of a different mental framework. Colonialism basically was patterned after the father-child relationship. These countries in the south were not mature yet. They were ignorant. They were somehow still in the childhood of human evolution and so had to be put under authority and under moral supervision. Lord Lugard, in the 1920s, concerning the British empire, has described the mission of England in a double way. He said, first the mission of England is to profit from the overseas territories, and second, and more importantly, it is necessary England be there in order to lead the natives on to a higher moral plane. Now it is after the Second World War that these two different goals converge into one, and this is called development, because now economic mobilization equals higher civilization. And this convergence is possible because now we don't talk about a relationship of authority between England or France and the overseas territories, but we talk about the relationships of commerce, of trade, of markets.

So on the one hand the mission of development was a mission which was gladly assumed and defined by the United States, which became the dominating world power after the Second World War. In order to project its global mission, it needed development, in particular, because development was not linked to the colonial discourse. However, it is also clear that after the war there was not only a world power seeking its mission, there were also many new states emerging, new governments being formed who were seeking a raison d'être as well. The anti-colonial movements in the 1950s and 1960s, and in particular the young nations, searched desperately for a justification of why they were there, which was not an easy thing to do because in many of these countries, states, at least modern states in our sense, were not known. So in order to impose taxes, in order to set up administrations, in order to extend control, in order to mobilize step by step a whole country, some mobilizing goal had to be set up. And this was development.

David Cayley
How do these nations consent to feel about themselves, in your view, when they agree to be portrayed as underdeveloped or developing?

Wolfgang Sachs
In the 1950s and the 1960s, most of these young nations set out to run a race that led toward a dead end. At that time it would have been possible to start off in various directions according to the heritage of each country, but many countries today have embarked upon the same path, and they are running in the same direction, and find it increasingly difficult to find their own way, to branch off from the direction that has been indicated by the United States. They find it difficult to invent, to create their own project as a society, to work for a way to live together, to produce, to be in a life that better conforms to their traditions, to their long-standing aspirations, to what they really are inside.

David Cayley
You spoke a little while ago of Truman and of the origins of this discourse, and the invention of the underdeveloped areas. You've traced the idea of development through a number of permutations. Can you outline these phases until you come to the present, to sustainable development?

Wolfgang Sachs
In the 1950s, development was basically the result of capital investment. So you would transfer capital, you would transfer later on certain qualifications, certain technical assistance. This

input was supposed to get development going, to reach that point of take-off beyond which, as it was said, development or growth would be self-sustained.

There were then modifications in the 1960s. One discovered that development is not only a problem of capital investment, it also has to do with people. So then one talked about manpower planning, schooling and education, in order to form manpower for development, to staff, if you want, the apparatus of producing a GNP. Then, by the end of the 1960s, a watershed was reached where it was recognized that transferring capital-forming manpower was not enough. On the contrary, many development efforts had produced quite unexpected results. With development, poverty grew, and people in the Third World didn't become richer or didn't somehow embark on a general upward street, but societies in the south polarized themselves. Some people became much richer and many others became much poorer.

That was most forcefully, at least from a prominent political stage, expressed by Robert McNamara in his famous speech in September, 1973, before the World Bank Assembly in Nairobi. He drew that conclusion saying, We have to acknowledge that poverty has increased, and that development leads to the rich getting richer and the poor getting poorer. Now this acknowledgement, this admission, if you want, did not lead to what you would expect, that one would have abolished or abandoned the politics of development because they had failed. No, it led to another operation. It led to the extension of the concept of development, and immediately McNamara introduced a new concept. He talked about rural development, about equitable development. So one dealt with the failure of development by extending the meaning of development. It was as if you had a building; and when you saw that the building didn't really fit, you added an annex to it, but the old building stayed there. You just added an annex to it — a second entrance, if you want.

David Cayley
This building of annex after annex is Sachs's paradigm for the history of development. Development theory, in his view, has become something like the Ptolemaic astronomy of the Middle Ages. One deals with the fact that the theory doesn't really

describe the motions of the planets all that accurately by constantly adding new epicycles to their orbits. Eventually the theory becomes meaningless.

Wolfgang Sachs
Whenever one noticed failure, destructive effects of development, the concept was extended. It exploded. And it ended up that the development included both the injury as well as the therapy. So it was development to bring big dams to India in order to increase the production of electricity, as it was development to heal the wounds by working with the tribals there, who had been driven out of their land because of this big dam. The injury was called development and the therapy was called development.

Consequently, development becomes a word which doesn't say anything anymore. It means one thing and it also means the contrary. This pattern has been maintained until recently and now, in the middle of the 1980s, the rise of sustainable development as the new catchword certainly signals a new age of development. Also, it was increasingly recognized that conventional development leads to environmental disruption. Now again, the consequence out of that recognition was not to finish with the business of development, but to extend development, to keep on a politics of growth. Maintain the conventional politics of development on the one hand, and on the other hand, now also take care of the environment, invent new methods to deal with problems of resource management, problems of environmental dislocation, and problems of pollution.

For that reason, then, the Brundtland report, which calls itself a report on environment, calls for a five-to-tenfold increase in the world GNP over the next twenty to thirty years. So again the same logic is at work. And in this case it is also nicely caught in the word itself. It is an oxymoron to talk about sustainable development, because, if you want development, then if it has in any way the same or similar meaning to what it used to be, it means non-sustainable. And if you want sustainability, it's very questionable that you can have development.

David Cayley
By the time we get to sustainable development, we're a long way down a chain of consequences, where each new phase of

development is cleaning up the last phase or absorbing some new contradiction into this growing amoeba, as you have called concepts like development. Even if one accepts that's true, however, by the time you get that far down the chain, is there an alternative?

Wolfgang Sachs
The Brundtland commissioners want to have their cake and eat it, too. On the one hand, they would more or less like to continue the politics of development and growth, which have been around since the time of Truman; and they would like to continue the enterprise to boost the GNP and to close the gap a bit between north and south by bringing the south closer to the north. On the other hand, they would like to do that in an ecologically peaceful way. I do not think both are possible. The perspective beyond the Brundtland report, then, would be to admit that the ideology of development today is obsolete, that it doesn't make sense to look for the future of southern countries by looking to the achievements of northern countries, and that only a politics of wide diversity, a politics of manifold experimentation, an attempt to find paths which grow out of the history and tradition of each country can, perhaps, — I don't say automatically — but can perhaps open up ways that make it possible to live in a decent manner on this planet, and to live in a decent manner without falling into the hands of a global, ecological management.

David Cayley
The alternative to development is usually portrayed as being stagnation. That is, one either develops or one underdevelops. You're saying that the opposite term is a culturally directed social project.

Wolfgang Sachs
First, there is not one development. There can only be many, many developments, but then it doesn't make sense to talk about development anymore. There are different, if you want, projects, ideas, directions to follow, different guiding images. And they have already been there.

Think of Zapata, who, early this century, led the Mexican peasants to revolution under the slogan and under the image of *ejidos*. The hope was to create *ejidos*, which means a certain form of collective agriculture, a certain way of independent, collective communities, based on the Indian tradition. So it was an idea of what the good life is about, that came out of Mexican history. The same was true for Gandhi. Gandhi's key word was *swaraj*, which meant a mixture of inner independence and outer independence. It was an idea that, again, had to do with the thousands and thousands of villages in India, which, to Gandhi, were looking for a way to remain as villages and to conform more to their own ideas.

The contrary to development is ... let's say global experimenting. For we are in a situation where the one royal path towards higher development doesn't exist anymore. Each country in the world is now faced with the question of where to go, and no country has a compelling answer. So all countries are in search. The only thing you can do is, if you want, broaden the possibilities, to let flourish what is there, to increase the richness of forms of life we have in this world.

David Cayley
It seems evident that the royal road is crumbling. As awareness of some kind of environmental crisis intensifies, it seems the dead end becomes more evident, but this doesn't necessarily mean that the lure of the modern then disappears, does it? It doesn't mean that everyone suddenly wakes up from the dream?

Wolfgang Sachs
It is clear that the most important effect of modern technology is a symbolic effect. Whatever we in the north have created has a tremendous impact on the imagination of the peoples of this world. Even if they have no means to live like us, their heads are full of images of our world; the images they used to have in their mind are fading away. So they are going to be stuck in that dilemma on the one hand, of having their minds set on the style of life in the north, which is projected to them in the idealized fashions of television, and on the other hand, of not having the possibilities or the resources to do that. How to get out of that impasse is a deep historic question, which will determine not only the end of this century, but also the next century.

David Cayley
So when you say that development is over, that you're doing an archaeology in a ruins, is this quite strictly true?

Wolfgang Sachs
Well, I do it in a polemical fashion. I don't make an empirical statement, but I would like to clear the possibility for a debate or a fight, if you want.

David Cayley
Wolfgang Sachs calls development an amoeba word, a plastic, verbal element which can be used to lend weight to statements that no longer have any precise meaning. As such, he considers that it's become a hindrance, rather than a spur, to creative thought.

David Brooks disagrees. He thinks that sustainable development can be given a precise, operational meaning. Brooks is a long-time environmentalist and a founder of the Canadian branch of Friends of the Earth, Energy Probe and the federal government's Office of Energy Conservation in the early 1970s. Today he's associate director for environmental policy in the social sciences division of the federal government's International Development Research Centre. We spoke in Ottawa and he told me that he thinks that the U.N.'s Brundtland Commission provides the necessary framework for rethinking development.

David Brooks
The Brundtland Report, for all of its deficiencies and fuzziness and wanting-to-eat-your-cake-and-have-it-too kinds of statements about development, was a pathbreaking document. The very fact that it was written by politicians and not by environmentalists, the fact that it was a consensus and not just an east-west consensus, which is turning out to be relatively easy, but a north-south consensus, made it important and, within the notions and the way it's been developed since then by both ecologists and economists and political scientists, we're getting to a framework, an operational framework that is something you can use to decide what you should do tomorrow and next year and the year after to make policy choices

that's far, far ahead of what was available at the time of "Limits to Growth."

When "Limits to Growth" was published by the Club of Rome in 1973, it was seminal — it gave us a term, it gave us a concept — but it was naïve in its general implications. In describing the policy conclusions of "The Limits to Growth," Dennis Meadows and others often use the metaphor of a ship sailing toward an iceberg. They would say that when the lookout sights an iceberg, the captain doesn't simply say, "Look, you're a worrymonger," and dismiss it. What the captain does is stop the ship.

I don't think that analogy makes any sense at all. A much better analogy: we're an airplane flying in some mountains and when our navigator spots a mountain directly ahead of us, there are a number of things we can do, but one of them is not to turn off the engines of the airplane. You don't stop the airplane in mid-air. Similarly, you just can't stop modern economies. You can adjust them here and there, you can gradually build in new goals, but you certainly don't just turn them off. The Brundtland Report clearly recognized that. I don't think they carried it far enough. I think they weren't adjusting the direction of the airplane adequately enough. It did provide the basis for saying we can make much more significant changes and in fact we will have to make more significant changes.

David Cayley
Can you sketch in for me what you think happened between the Stockholm conference, "Limits to Growth," era and the Brundtland Commission? How did the environment-development discussion evolve over that period, as you understand it?

David Brooks
Essentially what happened, what was formalized by the Brundtland Report was a shift of cause and effect. At the time of the Stockholm Meeting, the emphasis was on what the economy could do to the environment. That is, as you grew, you were going to have adverse effects on the environment. In effect, it was the formalization of the need for an environmental impact assessment. It doesn't ask much about what you're

doing. It says, whatever you're going to do, just do it better, from an environmental perspective.

By the time of the Brundtland Commission, for a variety of reasons, but in both developed and developing countries, we'd suddenly realized it's not the economy that's affecting the environment, it's the environment that's affecting the economy. We had flipped the whole thing around and people were realizing we had to ask not only how to make marginal adjustments in the system, but where was the system going? How big could it get?

David Cayley
What was it that had flipped the discussion around?

David Brooks
One was essentially the failure of development, the recognition in the poorest countries that they were limited by environment, that as their environment deteriorated, because of naïve attempts at development, they were in fact worse off. Second, the global issues had become more apparent. No one who was working on these issues could be unaware of the problems in the oceans, of the growing concentration of gases in the atmosphere, of the effects of desertification, deforestation. They were now global phenomena. There is simply not enough room in the available carbon-dioxide space, we might say, in the atmosphere, for developing nations. We have to reduce the amount of environmental space we're taking up in the world, in order to let developing nations take up a little bit of theirs. It's that kind of trade-off that is implicit in the notion of sustainable development.

David Cayley
You've said that sustainable development is a paradigm shift. Why a paradigm shift?

David Brooks
It's a paradigm shift because what is important is no longer economic growth, but development in the sense that Herman Daly uses the term, which is a realization of potential. It's a quality concept. In effect, we shift the emphasis from per capita gross domestic product or per capita income, to quality of life.

David Cayley
Have the big organizations that have adopted sustainable development in fact undergone a paradigm shift or have they simply put the concept on the letterhead?

David Brooks
Neither one nor the other. It would be saying too little to say they've just put it on the letterhead, but it would be saying too much to suggest it's a paradigm shift. No, I think what's happened is that they've caught up with 1972, or 1972 has caught up with them. Most of the organizations that are adopting sustainable development have added environment to what they otherwise would have done. They aren't questioning yet what the meaning of development is, but it would be a mistake to discount the changes that we are seeing. For me, they don't go far enough. For most environmentalists, they don't go far enough. Environmental impact assessment is a very important step and it'll move from projects to groups of projects, from country assessments to policy assessments, and at each stage we'll be bringing more and more in. You take those gains. You don't pretend that they're the answer, but they are very important. There are very few of these steps that are counterproductive.

David Cayley
There is another view, which I associate with a group that I would call conservative in this debate. The one who has been the most interesting to me, if not influential, is Ivan Illich, who defined development nearly thirty years ago as the war on subsistence. He set tradition and culture against development and saw development basically eroding culture, eroding people's capacity to cope and to deal with their environment as they have traditionally done so. Now for that point of view, I think sustainable development is counterproductive. It isn't just one inadequate step on the road to reform, but it's something completely different, something more sinister. It's a further colonization of culture by economics. How do you see that?

David Brooks
It's a very important perspective. It's one that I agree with on the one hand and don't agree with on the other hand. Certainly, the protection of subsistence options, the protection of cultural diversity, is critically important and there are many things we can do to promote it. The problem is we often don't have the land space. We often don't have the ecological room to do those things.

I would take the arguments of Illich very seriously. It means that you probably make development projects, even sustainable development, even what I would think of as good development, as small as you possibly can. You give as much of the control as possible to the community, but the community is not going to be uniform. I don't want to fall into the trap or be seen to fall into the trap of making rural life in villages seem like some kind of ideal.

I remember Ivan Illich's book extolling the virtues of the bicycle. I don't think he's travelled around Winnipeg very much in January. And similarly, life in many villages was pretty difficult. It was not easy, even for those groups, women and children, ethnic communities, that are often the focus of the objections to conventional development. The balance isn't easy and it's not simple and there are no general rules.

David Cayley
Let's take Vandana Shiva's work as an example. Shiva doesn't say that life was easy. What I understand her to say is that the alternative that was proferred to people was no alternative. It didn't actually exist, this neutral, secular, degendered space that was supposed to open up before people. In fact it was a mirage. So what happened, and I think this was also what Illich meant by calling development "a war on subsistence," he meant that the result would be what he called "modernized poverty." I think neither writer denies people the right to choose their path.

David Brooks
Yes, they're saying that the development that they were seeing not only restricted options for people, rather than opening them, but in fact it took the most vulnerable members of society and made them even more vulnerable. Again, when

you start to take sustainable development in its broader concept, not the one that focuses just on natural resources, but when you consider what has been good development, then the options for differences are there.

David Cayley
The question that I want to raise then is what is development? It seems to me that if you trace this term back, as writers like Wolfgang Sachs have done, it comes into general use in the late 1940s with Truman, who for the first time identifies virtually the entire non-European world as an underdeveloped area. Now there are certain assumptions that are made at that time and seem to become accepted almost overnight. But the main one, it seems to me, is that there is some universal, homogenous process that can be called development. This idea then goes through a series of permutations as it fails and is redefined and it fails and it's redefined, and it fails and it's redefined. A cynic would be inclined to say, Why not abandon this misbegotten attempt to postulate some universal, homogeneous process called development and recognize that it was a failure, that in fact we can't live without some culturally generated notion of the good which directs us?

David Brooks
A local, culturally sensitive notion of the good does not seem to be incompatible with sustainable development, as I understand it.

David Cayley
It may be incompatible with a hundred million dollars from the World Bank.

David Brooks
Absolutely. That's why I said, the World Bank hasn't accepted the paradigm. They're still doing the same projects, except they're smoothing out the edges with environment, avoiding the worst ill effects, but it's not sustainable development. I still don't think they value local knowledge. I mean, a hundred million dollars — just the scale drowns any local project. We should be dealing with projects of fifty thousand and a

hundred thousand and even CIDA can't deal with it. They want something in the millions.

The real danger of development is that it assumes a common set of goals, probably also a common set of processes, and it also suggests that western scientific notions will prove to be what everyone's been waiting for in the developing world. They just don't realize it yet, so we will bring it to them. Obviously, anyone who's been existing for hundreds of thousands of years has been living a form of sustainable development. Sustainable development is what was there. We're trying to get to a form of sustainable development that leaves what was there, but at a sufficiently higher income level. It may not be higher monetary income, but a sufficiently higher income to give people real opportunities in a world that is increasingly crowded, that is a world of interconnections, and those who suggest we should just withdraw from development or that everything can go on as before are forgetting, first, the numbers of people involved. Population is a problem. And second, how many of those people live in cities? So immediately they have broken the links of a self-sufficient, independently operating society. You can't just walk away from it at this point and I think one of the answers to the cynics is what happens then? Suppose we close down all the multilateral banks and the bi-lateral aid agencies? That action is a recipe for political and ultimately military conflict — initially political within and ultimately military between north and south.

David Cayley
Ending development doesn't necessarily mean ending everything that has been included in this term. It might mean trying to think of these things differently, with new words and new approaches. I think you're saying that's what you want to do under the banner of sustainable development and I'm asking, is that prudent when in fact you want to reject practically everything that has happened up till now under the name of development? Wouldn't you be more — I don't want to say honest — but more revealing of your intention if you spoke about something other than sustainable development? I mean you can say with Daly, "I don't mean growth by development. I mean realization of potential." But then, like Humpty

Dumpty, you're saying the words are going to mean what you want them to mean, but what they have meant is growth.

David Brooks
Development has, for many people, meant growth, but it's not what it really means. The people who use that word are the Humpty Dumpties. We're using the word in its real meaning and I don't deny that it's radical. That's what a paradigm shift implies. I would be quite happy if development budgets, obviously not IDRC's budget, but the real development budgets were cut substantially. I don't think we need to spend much on concrete and steel and big earth-moving equipment. That's where the problems have lain and yet there are good development projects.

I've seen good CUSO projects; I've seen good Oxfam projects. I remember some water-development projects that CARE Canada was doing. They don't even start building until they've been in the village for a year, until people know what they can do with water, and what they're going to do and where they want the lines to go, and what it is that water will do for them and how they'll manage the water system. All of that happens before you start building a water system for a village. I think CARE Canada's model was a three-year process for each village and it's only the middle year that actually involves pipes and a little bit of concrete and some water pumps. The rest of it is talking, discussing, and letting the community find out how it's going to run that thing.

And you can do much the same thing with local electrical. Electricity does not have to come in from wires from a central utility. We may be operating systems with a capacity of five to ten kilowatts just improving local industry, making it much more efficient. Now when I say efficiency, I'm introducing a western notion. I agree. I don't think that implies that you disparage local knowledge. It does imply that you are changing something. So the only alternative is to pretend that the rest of the world isn't changing and that you can isolate some fraction of it.

David Cayley
Wouldn't the alternative be to say that the rest of the world can change outside of our tutelage?

David Brooks
The rest of the world *is* changing.

David Cayley
Understood. What I was saying was let it change, but development generally implies tutelage by western agencies and governments.

David Brooks
Development implies providing options, providing alternatives, suggesting different ways of doing things and letting the communities decide which of those methods to build into their system of operating. It's much like growing trees. Within limits, there's no reason not to experiment with new kinds of seed, with new kinds of trees. There are reasons, for example, to import trees from other areas and try them out. What's a mistake is to try them out on a large scale and to design systems that always favour the richer farmers. You can think of systems that favour the poor farmers, systems that can be operated with minimal capital and with the labour inputs that they have, and look for those kinds of options. Look for options that are efficient at small-scale, not at large-scale systems that work well with more rather than less labour inputs. But those may be somewhat different from what has been there before and as long as the local community has the option to accept or reject them, they're worth talking about.

*** *** ***

It would be wrong to describe the difference between Wolfgang Sachs and David Brooks as a debate, since I interviewed them both separately and then, for my own purposes, placed the interviews side by side. Perhaps in conversation they would find more points of agreement than are evident here. Sachs does allow, at the end, that his stance is polemical; and Brooks, for his part, does not deny that he intends to reclaim development only by completely redefining it. Nevertheless, there probably is a real difference here.

The term *development* represents both an institutional and a conceptual matrix within which north-south relations have been conducted for the last forty years. In declaring this matrix obsolete, Sachs is trying to conjure away this heritage and open new horizons. For him, even sustainable development implies a single goal and single process for attaining it, and thus inhibits the "manifold experimentation," which he believes to be the only way out of the impasse to which development has led. For David Brooks, this view simply fails to take account of the world as it is: urbanized, overpopulated, its "ecological space" already contested and pushed past its limits, its peoples already committed to modernization. Under these circumstances it seems to him that redefining development, not rejecting it, is the most that can be hoped for.

In my conversation with David Brooks I quoted Humpty Dumpty's remark in Lewis Carroll's *Through the Looking Glass* that he could make his words mean precisely what he wanted them to mean, neither more nor less. It occurs to me now that, equally relevant to both thinkers, is the White Queen's remark to Alice in the same book that when she was Alice's age she was able to believe at least six impossible things before breakfast. The question is, who believes the impossible thing: Brooks with his assertion that development can be made good, or Sachs with his hope that it can be abandoned this late in the game?

13

Development and Democracy

"Development," Jane Jacobs once said, "can't be given." It's an organic expression of what a society is, and its preconditions are too complex and various to be conferred by one society on another. For forty years, nevertheless, Western societies have been trying to give development to the countries of Asia, Africa and Latin America. For the first half of this period, this project enjoyed tremendous prestige. The obligation of the rich nations to the poor nations was the moral pole star of international relations. Youth were conscripted for the crusade through CUSO and the Peace Corps. Then dissenting voices began to be heard. One of the first was Ivan Illich. He called development a war on subsistence, and predicted that it would undermine people's capacity to cope with their environments in traditional ways without offering a real alternative. The attempt to transplant western institutions, he said, would produce not western-style development, but social polarization with the majority forced into a situation of modernized poverty far more painful than traditional subsistence.

During the 1970s, parts of this critique began to be picked up by environmentalists. They noticed how big dams often displaced whole communities, how export-oriented agriculture stole the best lands from food production for local consumption, how commercial logging disrupted traditional harvesting of forests. Then in the 1980s, other critics began to notice how development was subverting democracy by putting money into the hands of unrepresentative élites, offering them, in effect, power without responsibility.

One of the most effective of these critics was Pat Adams. She's the executive director of Probe International, which describes itself as a public interest research group monitoring the effects of Canadian aid and trade policies on the people of the Third World. Probe International has been a thorn in the side of the national and multi-lateral development agencies and a voice for the disenfranchised victims of unwanted aid. In the fall of 1990, for example, Probe International gave External Affairs Minister Joe Clark a copy of a book it had just published called *Damming the Three Gorges: What the Dam Builders Don't Want You to Know*.

The book concerns a massive dam that the Chinese government proposes to build on the Yangtze River, a dam which will require the relocation of more than a million people. The fourteen-million dollar feasibility study for the project was financed by CIDA, the Canadian International Development Agency, supervised by the World Bank, and conducted in secret by a consortium of Canadian utilities and engineering consultants, including B.C. and Quebec Hydro. The consortium made a thirteen-volume study, which endorsed the project. Only a summary was released to the public. Probe International immediately petitioned for the release of the entire study and, after prolonged wrangling, received a somewhat censored version of it in April of 1989. Probe then invited ten internationally recognized experts to make an independent evaluation of the consortium study. This resulted in the volume forwarded to Joe Clark. It roundly condemned the feasibility study, both on the grounds of conflict of interest — members of the consortium would be in line for contracts if the dam proceeded — and of negligence in the assessment of the dam's human and environmental consequences. The consortium's study, said one of the book's contributors, Vaclav Smil of the University of Manitoba, was neither engineering nor science, but "an expert prostitute, paid for by Canadian taxpayers." "We regretfully conclude," said Probe International's director, Pat Adams, in her accompanying letter to Joe Clark, "that the Canadian government's commitment to sustainable development and to respect for the rights of Third World citizens is hollow."

The attack on the Three Gorges Dam, and on CIDA's role in bringing it nearer fruition, is typical of the work of Probe International. And, through this kind of work, Pat Adams says, the contradictions of foreign aid have been exposed.

Pat Adams
There has been quite a transition over the last ten years from a general acceptance and support for the development aid agencies and the concept of development in the Third World, to a more critical attitude. I would say there has been a loss of innocence. The public in the first world and in the donor countries in the developed world have started to see development for what it is. And one of the things that made the difference was that we started receiving some very nitty-gritty details of the implications, both environmental and human implications and consequences of these very large development projects, which were being designed in the capital cities of the industrial countries and in the borrowing Third World countries as well. And it was really this amassing of a huge amount of evidence of development projects that had gone awry. The reason that they are going awry is because they are not consistent with the wishes and the choices for the kind of lifestyles and use of physical resources that people in the Third World want to make.

We have to recognize that most Third World governments are not elected by their people and, therefore, when they choose a project, such as a hydro-electric dam or a road-building scheme, we should not automatically assume that that project is the choice of the people. In fact, we should assume the opposite because there are no checks and balances, or very few, in these countries to ensure that the projects that are chosen are really the choices of the people.

So it was really the collection of a huge amount of information and there was another significant thing that happened and that was the improvement in communication technology. More and more groups in the Third World, really citizens' rights groups, who were either defending communities because of the environmental consequences of a project that they were facing, or the social consequences or the economic consequences, would find us. They would organize themselves, they would try to fight these projects, and then they would say, now it's Canada that's financing this project. Who in Canada can help us? They, of course, realized that we were in part responsible for these projects and they found us. They found us through the churches, they found us through conferences, they found us through friends of theirs who happened to travel. They found whatever way they could to communicate with us. And for thirty-five years the development institutions, institutions like CIDA, the Canadian International Development Agency, the World Bank, have been able to spend money in other countries without us knowing what the consequences were, because we couldn't communicate with the people. That changed in the last decade. All of a sudden we started to get a lot of information about the consequences and we've started to realize that there are grass-roots citizens' groups all over the Third World that are just like the environmental groups and the citizens' groups in this country who are trying to defend their communities from unwanted development, from unwanted investments in the use of their resources. They want to make the choices, just as any Canadians would want to make the choices about how our own environment is used.

David Cayley
In 1985, Pat Adams and her colleague, Lawrence Solomon, brought out a book about what they were learning, called *In the Name of Progress: The Underside of Foreign Aid*. The book pointed overwhelmingly to the corrupting effects of development assistance and the way in which foreign sponsorship has allowed governments to ignore the wishes of their own people and commit follies they could never otherwise have afforded. Pat Adams has just published a second study along the same lines called *Odious Debts*.

Pat Adams
Gustavo Esteva, who spoke at a conference in Tunisia a couple of years ago, made a comment. He said, "In Mexico, we have really been enjoying the debt crisis." And of course that caught the attention of everybody and everyone's jaws dropped, but a number of us who were listening to him, who have worked

in the environmental movement and have been fighting a lot of these big aid and so-called development projects, understood what he was talking about. It struck a chord with us. He said that, essentially, money dried up when the debt crisis hit, banks were not prepared to lend more money to Third World governments, and that bought to a halt a number of very disastrous, so-called development projects.

For years the environmental community has been fighting ill-conceived, ill-considered development projects, such as hydro-electric dams in very sensitive areas that forcibly re-settle hundreds of thousands of people. The nuclear-power expansion programme in Mexico, for example, was also cancelled because of the debt crisis. They managed to finish two of their reactors, but just barely, and they cancelled a massive programme that they had planned on embarking on. Hydro-electric dams all over the world were put on hold; road-building schemes, logging operations were cancelled, because there wasn't enough money to finance them.

What Gustavo Esteva said was something that we had all recognized but hadn't really articulated, which was that money is power. And when you lend money, when your commercial banks or your government lends money to another government, it gives them the power to use resources in a certain way and it gives them a great deal of independence from their own people. And this is something that the astrophysicist, the well-known civil-rights leader from China, Fang Lizhi, said: that when his government, the Chinese government, is lent money from the outside world, it gives his government the power to be independent from the Chinese people. His government does not have to rely on the Chinese people in order to extract money to finance a lot of these big projects, and that's why he has called for a halt to all World Bank lending, all foreign-aid lending, except in the area of education, because he too articulated the same principle that Gustavo Esteva did and our colleagues in Brazil have articulated: money is power, and when we lend money to Third World governments, we give these governments power against their own people.

And for that reason, groups in Brazil, for example, have said to us, "Look, we appreciate your good will, we appreciate the principles and the concept behind foreign aid, we appreciate that you want to help us, but frankly, what you are doing is financing our government against us, so please just keep your money." This came as a real shock to us at the beginning of this decade, and over the decade we've seen how money can be misused, and this is really what has changed the attitude towards the development institution, the institution of channelling what is now forty-five billion dollars a year to the Third World.

David Cayley
You have just published a history of the debt crisis. How did the spending bonanza that led to the debt crisis come about in the first place?

Pat Adams
The main reason that is recognized by everybody is that there was a massive surplus of cash at the time of the oil crisis. What happened was the OPEC countries all of a sudden found themselves holding an awful lot of foreign exchange, which, of course, was being paid to them for their oil by countries like the U.S., all of western Europe, Japan, and of course the Third World. So there was this huge chunk of money that found its way into the hands of OPEC member countries. They wanted to do something with the money, so they deposited it in the commercial banking system. The commercial banks then found themselves with an awful lot of money, and they, when they accept money as deposits, then have to lend it out again, and so they did. They lent it out to the countries who were extremely short of foreign exchange and those were the countries in the Third World.

To what extent there was official encouragement from institutions like the World Bank, the U.S. government, the Canadian government, and western European governments, we don't really know for sure. It's very difficult to sort that out, although certainly bank presidents, such as Mr. Ritchie from the Bank of Nova Scotia, have said that there was sort of a wink-and-a-nod that the commercial banking system got from governments in the western countries to make sure that this money got into the hands of the countries who were very cash-short and those were the Third World countries. So they embarked on an extraordinary, massive lending programme, channelling billions of dollars to governments, many of which were military governments, almost none of which were elected by their people and over which the people in the Third World had no control. In Brazil, the Brazilian congress is now trying to track down the contracts for these loans that were made with commercial banks. They can't even get the contracts. You know, these governments were extremely unaccountable and of course the people in the Third World, if they dared ask the question, "How much money are you borrowing and what are you doing with it?" were very likely to end up in jail. So we can understand why they didn't ask those questions. They had no idea what money was being borrowed in their name, as it was.

David Cayley
And what turned this lending into what we now call the debt crisis? At what point did it begin to be perceived as a crisis and why?

Pat Adams
The crisis hit August 12, 1982, when the finance minister in Mexico phoned the head of the IMF and the Federal Reserve in the U.S., and said, "We're bankrupt. We can't pay our bills." And then the whole world really came crashing down. It was at that point that all of the banks realized how terribly vulnerable they were to a couple of countries in Latin America. They had lent out far more than they ever should have and their own financial viability was threatened if these countries could not continue to pay the money back. Indeed that's precisely what happened.

So a number of rescue operations were organized by the IMF and they managed to get enough cash back into the hands of the Brazilian government, the Mexican government, and the Argentinian government, so that they could continue to pay their bills. But at this point it was not new money. It was just new money being lent in order to pay back old bills. So it was just a very elaborate recycling process that was going on, designed to keep calm in the international financial markets.

David Cayley
Pat Adams considers the debt crisis an environmental boon, because it has slowed down big development projects. In seeing development itself as the primary cause of environmental destruction, she dissents from the current conventional wisdom of, for example, the Brundtland Commission. Brundtland argued that poverty is the main cause of environmental degradation and, therefore, only economic growth can save the environment. Adams disagrees.

Pat Adams
Most of the environmental damage that has been done in the Third World, especially in the last twenty years, has been caused by massive projects, such as hydro-electric dams, road-building schemes, cattle-ranching operations and agricultural schemes, that were financed with foreign money. But the development institutions like the World Bank and CIDA prefer to describe the environmental problem in the Third World as being a consequence of poverty.

I disagree with that. I don't think that the poor naturally destroy the environment. In fact, what has caused the destruction of the environment are very bad projects, and unsustainable economic policies. For example, in Brazil, there was a credit programme for the agricultural sector, which encouraged farmers to borrow money from the government at very low interest rates to purchase land with and then because of land-tenure regulations, clear the land in order to establish ownership of it, and most of that land was in the Amazon. This scheme was something that was very expensive to the government. They could only finance it as long as they received international financing, and it encouraged a massive destruction of the Amazon rainforest.

There were other projects as well. For example, the Balbina hydro-electric dam in Brazil, designed with help from a Canadian engineering firm, Montreal Engineering. They identified a site in the Amazon that has turned out to be a very bad site for a hydro-electric dam. It was not only a very flat area, but it turned out to have a couple of rather deep river valleys as well as ravines, which caused havoc when they eventually closed the floodgates and started to fill the reservoir.

Apparently what the engineers did was they flew over the area to be dammed and they used aerial photographs to measure the top of the rainforest canopy. They assumed that that measurement determined the topography underneath the forest canopy. Well, what they didn't know was that the top of the forest canopy was hiding a number of very deep ravines. When the time came and they closed the floodgates for the dam and the water started to back up and what it did was it created a series of canals and the water flowed into these little ravines and created a series, in fact, fifteen hundred hilltops and, of course, the water spread everywhere and it spread to a much larger area than they ever expected. Now on these hilltops the animals from the rainforest sought refuge and one of the residents described it as a most horrible scene. The animals who found themselves on these hilltops didn't have enough room and started to die and this resident described it as absolutely horrible and said that it was corpses on top of corpses on top of corpses.

But that was just the beginning. When they closed the floodgates, they had not cleared the area of the trees and the decomposing vegetation that was now under the reservoir caused the development of an oxygen deficiency, which led to the death of all the fish in the river. Not only did it do that, but it turned the water very acidic, so that the populations that lived around the reservoir now found themselves with intestinal disorders, skin rashes, vomiting, and there was a break-out of malaria, because often when you create very large bodies of water that are stagnant, then it creates a perfect breeding ground for mosquitoes, which are the vector for malaria. So the ultimate insult of this project was that it cannot generate the electricity that they expected it would generate in the first place, because there wasn't enough water to go through the turbines to generate the electricity.

Now the irony in all of this is that in 1985 an IMF, an International Monetary Fund austerity programme, and an austerity agreement that they had reached with the Brazilian government had led the Brazilian government to cancel the Balbina dam. The governor of the state of Amazonas, where the dam was located, objected strenuously, contacted the president of Brazil, and said, "I want this dam to go ahead," and there was, I gather, a great debate, and President Sarney eventually agreed that it would be exempt from the IMF austerity programme. The dam was completed in 1987. Now the IMF is the institution that the world loves to hate because it requires these austerity programmes from Third World governments. But austerity programmes are not always necessarily bad. In many cases, the money has been spent on projects that are not in any way suitable, either economically or environmentally, and I think we have to take that into account.

David Cayley
So you're saying that essentially the debt crisis and the drying up of new money stopped a lot of projects that would have been damaging. But isn't there another side to it, that the hardship imposed on these countries created other kinds of ecological problems?

Pat Adams
Yes. That is certainly true and there are cases, for example, in Ecuador, where the logging regulations were relaxed specifically so that more logs could be exported, so that more foreign exchange could be earned, so that they could continue to pay off their foreign debt. There's no doubt about it that the debt crisis has hurt the environment in some respects. Also in Ecuador, for example, oil exploration has been carried on at great cost to the tropical rainforest and to the native people who count that as their home. There's certainly no doubt about it. At the same time, I have discussed with Brazilian colleagues what would happen if there were no debt crisis, would the Carajas mining operation in the northeastern part of the Brazilian Amazon, which is a massively destructive mining operation that's destroying about fifty-eight thousand square kilometres of tropical rainforest, would a project like that not go ahead if there were no debt to be paid back? And they thought about it, and they said, "No, it probably would go ahead, because our government wants to export minerals, logs, whatever we can export in order to earn foreign exchange." So I think that there's always going to be the desire amongst several governments, all governments, for that matter, to earn foreign exchange, and they will sell off whatever they can of their country's assets in order to earn that foreign exchange. And the only way to restrict them from doing that

is really to empower the people whose resources are being pirated. That's the only thing that will stop governments and that's true of all governments, not just Third World governments, but our governments as well.

David Cayley
So you're saying for example in Africa or Central America, where lands have been taken for export crops and subsistence has been injured by that, that you think that probably would have happened anyway under the present political circumstances.

Pat Adams
Yes.

David Cayley
That it's not driven by the debt, it's drive by other forces.

Pat Adams
That's right. And it was happening before the debt. Indeed, that's where the debt came from. The money that was borrowed was borrowed to invest in hydro-electric dams, so they could mine the bauxite, so they could process it into aluminum, so they could export it for more foreign exchange. The problem came in that the money was borrowed for projects that were not carefully considered. You know, on paper they sort of looked okay, but we've managed to get a few of those papers in the last ten years, we've managed to get a few of those feasibility studies, and we've realized that the emperor has no clothes, that these projects never made sense from the beginning. And that's where the debt came from and we have to recognize that. If we turned the taps back on, if we could do away with the debt today and turn the tap of money back on, I can guarantee that in ten or twenty years there would be a new debt crisis, because that's where it came from.

David Cayley
A lot of people are calling for forgiving the debt as a way to get development back on track, the way to solve, say, the crisis of Africa. What do you think, in more detail, would be the consequences of debt forgiveness?

Pat Adams
I have no problem with debt forgiveness. I think that probably the vast majority of today's 1.3 trillion dollar debt that the Third World owes to us was contracted under fraudulent circumstances. And I think it is really outrageous that the people of the Third World were never party to these contracts, never had the means of due process to control their governments, and so on, before they embarked on these contracts. I think, in principle, for that reason, that they should not be expected to be responsible for these debts. However, if we did away with the current debt and could somehow conjure up lots of new money, as many of the development institutions are calling for, there are no guarantees that the money would be spent any better today than it was spent twenty years ago. A lot of the money would go into the same crazy projects that made no economic sense, that hurt the environment, that destroyed the environment of millions of people, and, ultimately, it is usually money that is lent — it's not a grant — and the people of the Third World would once again have to pay that money back. As long as there are no democratic checks and balances that a people can have over their government, there's no guarantee that our loans to them are either going to be properly spent or put in the right kinds of investments.

David Cayley
You've written about the debt crisis. I remember a piece in *The Globe and Mail*, maybe a year ago now, in which you spoke openly about the fact that the debt crisis may have done a lot of good, in a sense. It may have stopped a lot of bad things from happening in any event. I gather that you've created a lot of controversy by that stand. What's been the nature of the controversy?

Pat Adams
Well, the controversy has been, I think, unthinking, and I also think missed the point. The fact that the debt crisis has stopped ill-considered projects that destroy environments, destroy the livelihoods of people, is really a perverse consequence of the debt. To use the debt situation to control the expenditure and investment of money is a crazy way to do it. A far better way is to use democratic mechanisms to control the way govern-

ments borrow and spend money. But the fact remains when you have so many governments which are unaccountable to their people, cutting off money to them in fact restricts their ability to invest in damaging projects. Now that is not to defend the debt as a way of controlling these investments. It is a consequence. We just have to accept that that is the result. But there's a very important lesson in that and that is that money is power and that if we're going to lend money to governments, we have to say to those governments, "Demonstrate to us that this investment is consistent with the wishes of your people."

David Cayley
But how on earth could such an assurance ever be given or gained? I mean, what if the government of Brazil were contemplating lending money to Canada for the James Bay hydro-electric phase two? Would they judge that the people of Canada wish that to happen, or would they look at the fact that the people who are going to be flooded out don't wish it to happen?

Pat Adams
You raise a very important point. What we have to start by doing is not setting elaborate review procedures for these mega-projects, which go on in this country and in the Third World, but we have to start by recognizing the primary rights of people in all of our countries of the world. In the case of the James Bay, the Cree are participating in federal and provincial review procedures because they have to, not because they want to. As far as they're concerned, it is their land. And as far as they're concerned, this should not be a subject for discussion, as I understand their position. As far as they're concerned, they don't want the project to go ahead and therefore it should not be a matter for discussion by any review hearing.

We have to do the same thing for people in the Third World. We have to, for example, start by recognizing that you cannot forcibly resettle communities, ever. If a community wants to move and sell their land to a utility so that they can flood it with a hydro-electric dam, well, okay. But the point is that we have to recognize the land rights, we have to recognize the property rights of citizens all over the world. And when we start to recognize those rights, then they will start to protect the environment.

There's a very interesting and tragic case in India, a site called Singrauli, which is the site for what may be one of the world's largest energy and industrial plans. There are twelve open-pit coal mines, there are coal-fired electricity generating stations, and this industrial programme has essentially devastated the community. For generations it was a heavily forested area. There were lots of wild animals on which the people depended for their protein. It was a very prosperous farming community. And then along came the Indian government and its electric utility and said, "Well, we're going to create a reservoir here and then we're going to put all these coal-fired electricity generating stations around and put in all these open-pit coal mines," and within a decade the community has been destroyed, the area has been deforested, and it has now been described by the Indian press as equivalent to the lower circles of Dante's inferno.

What has essentially happened is that coal ash has spread around the community. It's landed on agricultural land, creating a sort of cement-like substance which made growing crops very difficult. The women and children in this community have been forced to take jobs with the electric utility and they work at half the state wages. They work for twelve hours a day, and the incidence of death from respiratory illnesses is extremely high. The canal that feeds the coal generating stations leaks, it's damaged agricultural lands. It has essentially been a disaster. It is, as the people there have described it, hell on earth.

The world talks with great fear about what are we going to do when India and when China start to use more and more coal to meet their energy needs. We discuss these countries as if they have this insatiable desire to exploit the global commons and to destroy the world's environment. Well, that's nonsense. We must always remember that when a project is going to harm the global environment, it's first going to harm a local environment. There are always people in a particular area who are going to be most threatened, and first threatened by that project. In the case of Singrali, we had a community of five hundred thousand people who had been shuttled around and resettled and sometimes resettled over and over again just to make way for these projects, who are now suffering from death and illnesses caused by the development scheme. They were the world's first line of defense and yet their rights were not recognized. They had no right to defend themselves. They had no right to say, "No, we don't want an open-pit coal mine here, we want to carry on with our agricultural lifestyles. No, we don't want a coal-fired generating system here, because that's going to give us respiratory illnesses," and so on. They didn't have the right to say that. They just had to live with it. And had they had the right, had their property rights been recognized, not only would their community have been saved, but the world's environment would have been better off. The Singrauli site is one of the largest point sources of CO_2 emissions in the world. And so I think we have to recognize that rather than there being a divergence of local interests from the global interest, in fact, there's a convergence. And if you give people at the local level the power and the tools to protect themselves and their communities, then the global economy and ecology will look after itself.

David Cayley
I believe you. But if you carry local autonomy to that point, would there be any economic development at all in the world? Can't you almost always identify somebody locally whose ox is being gored in any such development?

Pat Adams
No. Human beings and communities are always changing. People are naturally innovative, different communities change at different paces, some communities don't want to change at all. And we all know the Mennonite communities and so on and they make a conscious, collective decision not to change. And that's fine, they're perfectly within their rights to do that. And then there are other communities and individuals who may want to change. What societies have to do, and communities, is establish a decision-making procedure that protects the rights of each one of them individually and protects their rights as a community. When they have those tools, then some changes will occur, some innovations will be made. To my mind, that's what development is. Development is not when a government backs a corporation that comes in and says, "We

want to put in a coal plant here," or, "We want to put in a hydro-electric dam here and we have the right to make that decision, how to use your land." That's not development. That's assault, but that's not development. And that's what has been happening for the last thirty years in the case of the Third World. It's outsiders from Washington, from Ottawa, from the capital cities, who have been making decisions about how to use somebody's environment. Well, you can't expect accountable decision-making when the people who make the decisions don't have to live with the consequences, not only the physical consequences and the environmental consequences, but the financial consequences as well. It's not an accountable system.

There has been a total breakdown in accountability in the development process. Another example of that would be the case with Toronto garbage. As long as Toronto can find another community to take our problem, then there's no pressure on us to deal with our own problems, and as long as we can dump our radioactive waste on some other community, we never have to come to grips with the consequences of living with that. So that's why you have to set up accountable decisions. You know, take the case of Toronto garbage. I think if the city of Toronto said, "Okay, sorry, citizens, we're not going to pick up your garbage anymore," you would see a dramatic change overnight. You would see two and a half million citizens walking into supermarkets with plastic bags or old yoghurt containers and saying, "I'm sorry, I've brought my own containers. I don't want to purchase the stuff in all this packaging." You would see a dramatic change very quickly, because we would be forced to live with our own garbage. And when you're forced to live with your own mistakes and your own problems, then all of a sudden there's this wonderful innovation and people find solutions. People will always find solutions. It's just that we have to have limits. We have to have limits placed on our activities and we have to have the rights of other communities respected.

David Cayley
This view seems to have gained you a reputation as a right-winger. Why do you think that is? Is it because you see solutions in law, in property rights, in well-established institutions,

rather than through the creation of new service bureaucracies to address new environmental needs? It seems to me your solutions are always essentially simple and already available, although very radical in what they imply, and that we are now on the threshold of an era of environmental services, where development enters a whole new phase. It seems to me that you're fundamentally going against the grain with what you're saying.

Pat Adams
Our solutions are essentially decentralized. Our solutions are to put power into the hands of individuals and individuals as they want to organize themselves into communities. And when you do that you have to give up power. You have to say, "I don't want a central government making these decisions. I want local communities making these decisions." And that is in some cases not consistent with the conventional left wing. It's not autarchy, it's really decentralized decision-making and I think it's based on a respect for the good judgement of the average person. But in order to accept that, you have to accept that power is going to be devolved from a central government or a central body. I think that frightens a lot of people. I think a lot of people feel as Eugene Black, who was an early president of the World Bank, felt, that the average person cannot make good decisions themselves, that there have to be these development diplomats and that they have to align themselves with the élites in Third World countries, because only they can figure out what is best for the people of that country. And to save the people of that country from themselves, they're going to have these experts making decisions for them.

What we're saying is that the best expert is the person who has to live with the consequences of a decision and once you make those people accountable and you also give them the tools available, make them available to them, to develop their own communities themselves, then you will start to have sound decisions. You know, I think a pulp mill, for example, which wants to establish itself on a river will have to seek the approval of all the people in that river basin who are going to be affected by that pulp mill and I dare say that they will have great difficulty doing it.

David Cayley
Do you think that there could be a pulp mill under your scenario?

Pat Adams
Under the current circumstances, I doubt it, because I don't think the technology has been developed to do it, but once corporations know their limits, once they know that they're not going to be able to put these projects in place because it's not going to be acceptable to the local community, they will find alternatives. But as long as they're not obliged to find alternatives, they won't. As long as the people of Toronto can push their garbage onto somebody else, we will. It's these limits which have to be established. We have to lose our ability to create poisons for other people.

David Cayley
I like that way of putting it very much. You've been involved with a whole campaign to identify the human and the ecological costs of big development projects, and I suppose when the World Bank cancelled the so-called second power sector loan to Brazil, that was at least a symbolic moment in which you won an important victory. The bank itself acknowledged the case against the big hydro-electric scheme as it then existed in Brazil. What can happen now? Is a green World Bank a contradiction in terms? What can come out of this conjuncture?

Pat Adams
The World Bank is doing a very good job of painting itself green. They are churning out an awful lot of rhetoric and they now are embarking on what they call the green fund, which is going to be about four hundred million dollars, from which they will fund so-called ecological projects. Now I think that it is impossible for the World Bank to be a green institution because the World Bank is a multi-lateral institution, is accountable to the people in no one country.

For example, if they propose a project, say it's a hydro-electric dam, they may or may not do an environmental assessment. If they do one, they are not obliged to release it to public or to peers for peer scrutiny. Now that means they can get away with murder. They can get away with whatever they want. They can say, "Oh, we've done an environmental assessment and we have decided after doing this assessment that this project is sound and we're going to go ahead with it." Not only that, but they are not obliged to discuss the issue with the people who are going to be affected by the project and certainly they're a long way from ever giving these people a prior right to make the decision. So there's no way to make sure that their wishes are consistent with the choices and the wishes of the people in the Third World.

David Cayley
The green fund that you mentioned, what would constitute an ecological project, as you imagine it? What potentially would be the uses of this money?

Pat Adams
One potential use would be reforestation and that, of course, is very dangerous because, whether you cut a tree or you plant a tree, you are affecting somebody's environment, and as one of our colleagues in India, Anil Agarwal, who's a well-known environmentalist there, has said the aid institutions are very good at planting the wrong trees in the wrong places for the wrong reasons for the wrong people. And the most common reforestation programmes are these massive eucalyptus monocultures which are popping up all over India, places like Thailand as well. Eucalyptus is not only very environmentally destructive, it actually absorbs a great deal of water, tends to really deprive agricultural communities of available water, chokes out agricultural crops, is not available, not acceptable to animals so you can't use it as forage, and so on. It's a very quick-growing tree and it's been primarily grown for the cellulose.

So planting a tree can be just as damaging as cutting down a tree and whenever you tamper with somebody's environment, whether you call it a green project or an environmental project or not, is irrelevant. The point is the people who are going to be affected have to be able to decide yes or no. Do they want their environment to be used in this way?

David Cayley
I have a feeling that in your ideal commonwealth there is simply no room for this huge, international bureaucracy, that they can by definition do no good.

Pat Adams
Well, there are a bunch of rules that they should adhere to and I think it's going to be difficult for them to adhere to them, but I think, if they want to continue to exist, they have to, otherwise they're going to continue to finance very destructive projects at the expense of millions of people and at the expense of their treasuries, which is very dangerous. I think the international institutions look at public opinion, and they say, "Ah, public opinion is green now. Therefore, how can we be green?" And the only thing they know how to do is spend money, but as long as they try to spend money in ways that are not accountable to the people who are affected, it doesn't matter how green they try to make it.

The World Bank is afraid to talk to people. Why are they afraid to talk to people? Their job is to improve the lives of people in the Third World. Well then, what are they afraid of? Well, they're a multilateral institution. They're made up of governments. Half of their members are governments that don't represent their people and therefore the World Bank as an institution can't go in and talk directly to the people. Well, if that's the case, then we're going to have to close the institution down, because if you're spending money that influences the way other people live, then you've got to be accountable to them. You've got to give those people the right to say no and if they don't have the right to say no, then mistakes are going to be made — a lot of mistakes with very, very large consequences.

*** *** ***

Pat Adams discerns in every environmental problem, no matter how global its manifestations, an original issue of local sovereignty. The Singrauli smelter complex in India may form part of the global problem of CO_2 emissions, but it first involved the disinheriting and disenfranchisement of the local people whose lands and livelihood were ruined by it. Viewing things this way turns vast, abstract and intractable problems into small, concrete and manageable questions of democratic rights.

Adams approaches the international network of development banks and aid agencies in the same spirit. She asks not whether the World Bank has added an environmental department and promulgated new environmental guidelines for projects, but whether there is any means by which such an agency can be made accountable. She concludes that there is not, and that no amount of good intentions in high places will prevent such a huge concentration of unchecked power from doing untold damage to virtually defenceless communities.

This approach offers a corrective to accounts of the environmental crisis which locate the problem in the nebulous dimension of attitudes to nature. Environmental damage, from Adams's point of view, is almost always the effect of ungoverned power. The remedy is as obvious as it is difficult. Meanwhile, Pat Adams, at least, continues to tell the truth about it.

14

The Steady State

Let's build for economic society the largest feasible playpen in which you can leave the child, the person, to be free. If we run around always trying to correct individual behaviour every time somebody does something a little wrong, we'll go nuts. So just build an area within which the market and people can be free, but set the boundaries so that we can't hurt ourselves by overstepping and destroying the ecological niche in which we live and on which we depend. If we can set those boundaries, we can rely on the market within those boundaries. But the market itself can't set the boundaries for itself. It needs collective, social, community action and coherence to set those overriding limits.

– Herman Daly

In 1987 the U.N.'s Brundtland Commission put a new term into general circulation: *sustainable development*. The concept instantly caught fire. Our prime minister committed his government to the idea, the Canadian International Development Agency made it their policy, as did the World Bank. The trouble was no one really knew what the term meant. The Brundtland Commission made a stab at a definition. They defined as sustainable, "development which meets the needs of the present generation without compromising the ability of future generations to meet their needs." But this only transformed the vexing problem of how to define sustainability into the even more vexing problem of how to define needs.

If the problem of what sustainable development is has a solution, it may rest on the work of economist Herman Daly. For twenty years, Daly has been grappling with both the theoretical and practical questions that arise in trying to define sustainability. What is the proper scale of economic activity for a given society? How can ecological limitations be incorporated into economic theory? How can societies gain the benefits of free markets without being destroyed by their unwanted side effects?

"There is something fundamentally wrong," Herman Daly once said, "in treating the earth as a business in liquidation." And yet, this is precisely how modern society has viewed the earth, as something to be used up, rather than husbanded or sustained. Endless growth and ever-increasing consumption are fundamental postulates of our economic science. The limitations of this approach are now widely evident. Adam Smith's invisible hand has become an invisible foot, Daly has said, inexorably kicking nature and society to pieces. But conventional economic theories can't come to grips with the problem. For this reason, Daly believes, we need a new economics. He calls his preferred approach a steady-state economics and he brought out his first book about it in 1971. There he posed the fundamental questions that have dominated his work. If economic society is to stop growing, at what level should it maintain itself and how is this to be accomplished? In 1989, Daly brought out an ambitious new book called *For the Common Good: Redirecting the Economy Toward Community, the Environment, and a Sustainable Future*. The book was co-authored by theologian John Cobb. Daly and Cobb argue that we have now entered the era of what they call "uneconomic growth," a growth that impoverishes rather than enriches, where "The faster we run, the behinder we get," as Alice says in *Through the Looking Glass*. They called for new, more sensitive measures of economic welfare, for a new economic anthropology which replaces the isolated human atoms of classical theory with the model of what they call "persons in community" and for a new sense of the absolute natural limits of economic activity. For many years, Herman Daly was a teacher at Louisiana State University. Today he's a senior economist in the environment department of the World Bank in Washington.

Herman Daly
Economic theory developed at a time when the human scale in the total biosphere was relatively small and so it made a certain amount of sense to think of sources of raw material and sinks for waste material as infinite, or not really scarce. So economics abstracts from whatever is not scarce. And now the scale of the human economy has grown to where it's no longer a negligible force in the biosphere, so we have to change that assumption. We also have to change our vision of the economic process as an isolated, circular flow, a commodities exchange from firms to households, and from households to firms. In this vision, the economy goes around and around in a circle with no inputs from the outside, no outputs to the outside. So this vision, which you find in the first pages of any standard textbook, is that of an isolated system. An isolated system has no environment. It has no points of contact with the environment.

So from the beginning you abstracted from these kinds of things and it's hard to bring them back in after you've developed your whole theory on the basis of this abstraction. When the environment comes along and hits you in the face, you have to deal with it as an "externality," which is why you see that word everywhere in economic literature nowadays — externalities, externalities. These are things that didn't really fit in the theory but they're too important to ignore, so we have to deal with them somehow. So you do it in a kind of ad hoc fashion as externalities.

David Cayley
Is it also a question of displacing problems onto nature to avoid human rivalry?

Herman Daly
Oh, yes. I think if you look at one of the main reasons for growth it's to have more for everyone, so you don't have to share, because sharing brings about conflict and people don't want to give up anything. You can just increase the total

amount. And that means encroach more on the natural world, take in more materials and energy to divide up among people. The big thrust for growth is to avoid sharing, or to put it off for as long as possible.

David Cayley
In their book *For the Common Good*, Herman Daly and John Cobb accuse conventional economic theory of what they call "the fallacy of misplaced concreteness." The phrase comes from John Cobb's mentor, Alfred North Whitehead. It means essentially forgetting that abstractions are abstractions and reading them back into reality as if they themselves were something real. Daly sees this mistake as endemic to the discipline of economics.

Herman Daly
Let's suppose that economic activity were making a pizza, then economic theory would abstract from the pizza one or two characteristics, in this case by analogy it would probably abstract the circular shape of the pizza and then economic theory would consist of statements about how fast the radius has to grow in order for the area of the pizza to double in a certain time, and then it reasons in these categories. And then along comes someone and asks, "Hey, what about cheese and anchovies and how much pizza is really enough?" And these questions are not easily answered in the context of the geometrical shape of a pizza.

David Cayley
The anchovies are an externality.

Herman Daly
They're an externality or they're kind of left out. So if you then draw conclusions about real-world economic pizza-making from this theory, which only looks at the circular shape of the pizza, then that's going to be fallacious. You're going to say, "Well, we can have so many pizzas this size," and it'll turn out there won't be enough anchovies or something. You will have abstracted from all of these other things, like anchovies and cheese, and reasoned only in terms of the circular shape of the pizza. That's the fallacy of misplaced concreteness. You've left

those things out of your basic abstractions and then you draw conclusions that have to do with those things and frequently they're wrong.

David Cayley
Does the fact that economics presumes first to be a social science and then to be an independent discipline also enter into this problem of abstraction?

Herman Daly
Oh yes. Abstraction is rampant, of course, in all disciplines, but it reaches toxic levels in economics. And of course I'm not knocking abstraction because it's necessary for thinking and, indeed, the person who identified the fallacy of misplaced concreteness was Alfred North Whitehead, who was certainly no stranger to abstract thought. But precisely because he was so adept at abstract thought, he recognized its limitations as well as its powers and emphasized them. Often in economics and in other disciplines we see only the power of abstract thought, which is considerable, and tend to be a little bit blind to the limitations that it raises. By the nature of the case, if you've abstracted from certain things that are important, then they're going to come back and haunt you at a later stage of your thought and policy.

David Cayley
Before we come to those accumulating externalities, what are the other major assumptions of classical economics you'd want to identify here as now being problematic?

Herman Daly
I think there's the *homo economicus* as the pure individual whose identity is totally self-contained so that all relations between individuals are purely external, that is, that the individual is defined independently of all his relationships with everyone else and all relations are external. John Cobb and I have argued that a better model of a truer *homo economicus* is that people are persons in community, that is, their very self-identity is made up of the most important of these relationships, or relationships are internal to the very definition of the individual, and not just external things to the individual. And

if you take that point of view, then community becomes important. Community is built into the definition of the individual and to what the individual wants and how he acts. In current economics, community is nothing other than the subtotal of individual relationships and all of these relationships are external, so community is just an aggregate of individuals. But we say community is much more than an aggregate of individuals. Community enters into the very definition of what the individual is, how he sees himself. If I am son of Mildred and Edward, father of Karen, husband to Marcia, you take all those things away from me, then there's not a whole lot left. There's something left, but it's not much. And so we say that all these relationships constitute the individual's identity.

David Cayley
You spoke earlier of the externalities that the theory doesn't take account of, which have to be added. And these accumulate. More and more are identified, which corresponds to Thomas Kuhn's theory of what happens to a scientific paradigm. More and more things are seen to be wrong with it. Do we need a new paradigm?

Herman Daly
I think so. And the externalities are almost perfectly analogous to Ptolemaic epicycles. You know, before Copernicus, and I guess Kepler, they always wanted to explain the motions of the heavenly bodies in terms of circles because obviously a circle is the perfect figure. I mean why would God do anything so weird as an ellipse? So you keep having circles and then circles piled on circles until you manage to trace out the actual pattern. And it worked, it was just terribly complicated. And then once the idea came, well, let's use an ellipse, then the whole thing was greatly simplified. So I think that's what we're doing with externalities. Externalities are epicycles. We just keep piling up more and more, and we need to recognize then that economic commodities don't just flow in nice little circles within the economy, but they take broad elliptical orbits which go through the ecosystem and then back through the economy and affect many different things.

David Cayley
You also expressed in your book reservations, however, about the idea of a paradigm, reservations about a powerful, new explanatory framework.

Herman Daly
Well, we've expressed reservations about a deductive system, because economics has modelled itself after physics and it wants to be a deductive system. Just get a few first principles that are right and then by mathematics you can work out everything else. Yes, we have a lot of doubts about that. We think that we ought to be much more historical and recognize the changing circumstance of time and place and not try to be so all-embracing in terms of economic theory.

David Cayley
Economics more as a sort of natural history of human activity?

Herman Daly
Yes, that's a good way to put it. It's more on the order of natural history, which has a few principles. It's not bereft of any unifying principles, but it tends to be rather historical in particular, rather than just deductive and totally general in its approach. And on the thing of paradigm, I've recently discovered that the great economist Joseph Schumpeter, long before Thomas Kuhn, had expressed the same idea using a different term. He spoke of a pre-analytic vision. He said before analysis can begin, you have to have a pre-analytic cognitive act, which he called vision, which gives you the basic shape of the thing that you're going to analyze, and anything that's left out of your pre-analytic vision can't be corrected by later analysis. That's sort of related to that fallacy of misplaced concreteness once again. Once you've left it out, you're not going to get it in. So I felt that Schumpeter's way of looking at things was very much like Kuhn's and I even think that the term "pre-analytic vision" is more descriptive than paradigm. I think that really tells you what's going on.

David Cayley
Herman Daly is seeking a new paradigm or pre-analytic vision for economics, but there's a great deal in economics that he

wants to preserve. In terms of Ivan Illich's distinction between those seeking an alternative economics and those seeking an alternative to economics, Daly is definitely on the side of alternative economics. He doesn't challenge the basic postulate of scarcity, for example. His aim is not to redefine economic activity, but to establish it within its proper limits, to see, for example, what markets are good for and what they aren't good for.

Herman Daly
The thing that economics does rather well, or at least markets do rather well, is to allocate resources among alternative uses by decentralized decision-making. Markets are a way of getting decisions down to the local level of the individual user and avoiding huge bureaucracies and central planning and all of that. That's what market economics does well. What it does poorly or what the market has no real capacity for sensing are two things: the first is the distribution of income. Markets will distribute income in a way that may be efficient from an incentive point of view, but it can be highly unjust. So the problem of justice in distribution has long been recognized. The second is the question of the optimal or proper scale of the entire human economy relative to the ecosystem. The market has a tendency to grow and so reaches a point at which the extra costs of further growth are greater than the extra benefits. Beyond that point, further growth doesn't make you richer, it makes you poorer, because it increases costs faster than benefits.

We haven't recognized this idea yet in our public policy. We generally take it for granted that the benefits of growth far outweigh the costs. We reason that they did in the past, so why won't they in the future? Well, they won't in the future because we're at a much larger scale now and we cause much greater impacts on the natural world, which produce much greater feedbacks from the natural world: ozone depletion, CO_2, greenhouse gases, acid rain. All these things are products of a large-scale intervention by human beings in the ecosystem and their costs are increasing faster at the margin than the benefits. To take an extreme example: If ozone depletion results from CFC propellants getting into the atmosphere, what's the benefit of these propellants? Instead of a finger pump on a can, you have a pressurized spray. Maybe CFC has some advantages in air conditioning. Okay, what are the costs? The costs may be increased incidence of skin cancer or the disruption of agriculture worldwide. At the margin, then, the costs seem to go up faster than the benefits in many dimensions of economic growth. We have to recognize the concept of an optimal scale of the entire human economy relative to the ecosystem, along with an optimal allocation of resources, and a just distribution of wealth.

David Cayley
Do we need to recognize this concept only because physical nature begins to kick back and there's a hole in the ozone layer, or do you think that human nature also revolts against the scale of economic activity that we presently have?

Herman Daly
Yes. I think human nature also suffers under this. This is partly what John Cobb and I were trying to get at with the idea of community, that the most satisfying relations people have stem from community and from some sense of belonging to a place and time and group with satisfying personal relationships, more than consumption of another tennis racquet or a golf club or something. So if in our striving for efficiency to produce more golf clubs, we end up destroying communities so that you can't find a golfing partner anymore, or that it's hard for you to make friends or talk to anyone, then we've given up more than we've gained and this is what John Cobb and I were trying to get back into economics.

David Cayley
How to get from here to there is something you've been thinking about for many, many years. And in a number of books, like *Steady-State Economics*, you've made proposals. Can we talk about what your major proposals are, first, to find the optimum scale? How can one think of that?

Herman Daly
Of course, one of our problems is that we don't measure the costs of growth. We just have the GNP, which is a mixture of costs and benefits insofar as they cause expenditures. We just throw them all together. We should separate out the cost com-

ponent of GNP and the benefit component, keep separate accounts and occasionally compare them, instead of just adding them together.

One thing John and I did in *For the Common Good* was an appendix, in which we developed an index of sustainable economic welfare. One way of interpreting that index is to say that what we found was that for the United States from 1970 to 1986, which was the last year of our series, extra costs of economic growth in the U.S. were sufficient to outweigh extra benefits, so that aggregate welfare was pretty much constant, declined even a little bit, according to what we consider to be a fairly reasonable measure.

David Cayley
Could you give an example of something that looked like a benefit when it was aggregated in GNP, which was actually a cost when you teased it out?

Herman Daly
There are two major categories that I found. One is the liquidation of natural capital — the forests, mines. You cutdown a forest beyond its natural regenerative capacity, then that's consumption of capital. The forest kept in its original state would yield a certain income, a certain sustainable yield of trees, year after year. But if you cut down the whole forest and in the year you cut it down you treat all that as income, that's not proper. That's capital consumption. That's like selling your house and spending all the money this year and thinking you're rich, but then the next year you're poor. Take mines as another example. When you're depleting a mine, you count all the copper sold in the current year as pure income, when a large part of that is capital. So, consuming capital. And the other thing is not subtracting what economists call regrettably necessary defensive expenditures, the expenditures that we have to make to protect ourselves from the side effects of other production, so that if a firm is polluting the air and causing medical upper-respiratory problems and you have to go to the doctor, then those medical expenditures are really the costs of producing whatever was being produced that caused you to get sick.

David Cayley
So according to your calculations, the U.S. has already entered an era of diseconomic growth?

Herman Daly
Yes, exactly. Uneconomic, or I'd even say anti-economic growth in the sense that if we grow now, costs seem to be going up faster than benefits, so that makes us poor. And it's hard for people to become accustomed to that watershed. Frequently people say, "Oh, we have to grow more, in order to be able to afford the costs of cleaning it up and of helping the poor." Well, nobody doubts that if you're truly richer then it's easier to do everything, including clean up the costs of growth, and help the poor. But the question at issue isn't that. The question is, "Does growth from the present margin really make you richer or is it making you poorer?" If it's making you poor, then we can't appeal to growth as a way to help clean up or pay extra expenses. It just makes things more difficult.

David Cayley
How long have we already been in this era of diseconomic growth by your reckoning?

Herman Daly
It's hard to say, but our rough calculations show that from 1970 to 1986, at least over that period, it looks like welfare in the U.S. has been pretty flat. Now, this depends on a number of assumptions that one makes in measuring the index, one of which in our case was that we weighted extra income to poor people more heavily than extra income to rich people and there are very good reasons in economic theory for doing that. There's the idea of diminishing marginal utility — everyone satisfies their most pressing wants first — and so a dollar income to a poor person means more food, clothing, shelter, basic needs. An extra dollar income to a rich person may mean a third TV set in a summer home or something, which doesn't really add to his well-being nearly as much as an extra amount of food adds to the well-being of a poor person.

David Cayley
To achieve a steady state, you've said first you would need the means of identifying what's going on in the economy and what are the benefits and what are the advantages. What would be the other prerequisites of a steady state?

Herman Daly
You need to have measures of costs and benefits to choose an optimal level at which to maintain a steady state, but we could maintain a steady state at various levels and not the optimum one. Basically, you need to limit two things: You need to limit human population growth and you need to limit the growth of per capita human consumption. So if we put some limits on reproduction and some limits on our per capita consumption, that's what I think is required. I think the easiest way to limit consumption, and I mean here consumption of resources, material, physical things, is to do it at the depletion end, at the input end, to restrict the amount that we extract from nature and bring into the economy. By restricting that, we will ultimately also limit the amount that exits as waste later on.

What we need, then, is something on the order of a depletion quota or a severance tax. As time goes on I tend to make more and more modest proposals as my more radical ones are ignored. So now, for the United States at least, I'm saying, "Here's a proposal which wouldn't get us all the way to a steady state, but I think it would be a nice step forward for the U.S. to put a heavy severance tax on resources, particularly energy." Raise most of our public revenue from a heavy tax on resources. Then ease up on the income tax, particularly the taxation of lower incomes and perhaps even have a negative income tax at the very low levels of income, again financed by receipts from the severance tax. This would do several things. It would limit the material inflow of resources out of nature into the economy. We've now made that expensive. Also, it would incentivate the technologies which would use these resources much more efficiently and productively because they're more expensive, so we're going to economize more on them. Just like we did when the Arabs raised the price of oil, we learned to be much more efficient with oil, but instead of paying the Arabs the extra money, why don't we pay it into the U.S. Treasury and ease up on income taxes and let the poor have some benefit as a result? In addition, a severance tax is easier to administer and collect than an income tax. It's harder to avoid it and it has less of an effect on incentives to work.

Here, then, would be something that we could do which would increase the efficiency of resource use. The technical optimists tell us that we can increase efficiency by factors of ten or twenty. Okay, if that's right, then let's do it. Here's a way to push. Pessimists say, "Well, we probably can't do that but we really need to limit the resource throughput." Well, this limits the resource throughput, so both optimists and pessimists ought to be happy with such a proposal. Well, that's kind of one step towards a steady-state system.

David Cayley
Don't you run into the problem if you're making the proposal for the United States of the interdependence of the United States? American producers end up competing with producers in other countries which haven't done this?

Herman Daly
Yes, this is a big problem. You, in Canada, have already had a debate on free trade, but in the United States it hasn't been debated and the big problem, the big conflict, is just as you've outlined. All economists will agree that the way to deal with environmental problems is to internalize the environmental costs into prices. If a country does that, then its prices go up. If its prices go up, it will be at a trading disadvantage relative to countries who have not internalized those prices.

Consequently, the internal policy of sustainable development or steady state is going to require some kind of external protection. You'll probably have to have tariffs to protect producers against countries who do not internalize those costs. Maybe when a whole bunch of countries adopt the same national rules of cost internalization, then you might have free trade among people who play by the same rules, but certainly you can't have some countries internalizing costs while others don't and then have them trade freely with each other.

David Cayley
So in other words, regionalization of economies, if not nationalization, making them more self-sufficient units, would be a

prerequisite for a steady state. Unless you could devise a way for everyone in the world to move at once, which doesn't sound too promising.

Herman Daly
Yes, that sounds kind of hard to do. So you're better off to have a greater degree of self-sufficiency. John Cobb and I argue for this in our book. We go back to John Maynard Keynes. It comes as a surprise to many people to find that he argued very forcefully for national self-sufficiency, not in the extreme autarchic sense, but he said something to the effect that ideas, art, hospitality, and travel are things that are international, but let goods be homespun as much as possible and also finance should be primarily national. We call that short-supply lines. Try to keep your supply lines short. We don't try to make a fetish out of being totally self-sufficient in any arbitrarily defined area, but just, other things being equal, keep those supply lines short.

David Cayley
It certainly doesn't seem to be the way things are going at the moment.

Herman Daly
No, it doesn't. There's a tendency to rejoice in the maximum possible interdependence and lengthening of supply lines. People seem to think that this ties the world together into one complex, interdependent unit, and that, therefore, people will all learn how to get along with each other because the cost of not doing so would be too great. But history does not support this theory. It just means that when we make mistakes, the costs in one area are going to be spread all over.

David Cayley
This may come from an earlier period when you were making more radical proposals that no one listened to, but in your first version of *The Steady State*, you also proposed maximum and minimum incomes to limit consumption.

Herman Daly
Yes. That was the notion and I still like this proposal. We didn't really make it in the new book but we came pretty close. The approach we take is limits to inequality. The idea is not to push towards equality because there are many good reasons for having different incomes, but unlimited inequality is a violation of community. If one person owns everything and everybody else owns nothing, then surely you can't talk about community. There has to be some sort of limited inequality that goes along with community.

If you go all the way to pure, to absolute equality, that's a denial of individual differences in the community, which we think should be respected. So there's some limited band of inequality. What should that range be? That's an empirical question. We can experiment. My own view is a factor of ten difference between the highest and lowest is enough to reward all important differences and still create a sense of community in which people respect the differences and need for rewarding greater efforts. Just look at a university or look at the military or the civil service, you generally find a factor of ten difference. I don't see any reason why it needs to be much greater than that. But what do we have today? It must be a factor of a hundred or more.

David Cayley
In other words, where people have to come to grips with this in a bureaucracy, they've already arrived roughly at this factor of ten difference.

Herman Daly
Yes. People ask, what will happen when people reach the maximum limit? There's a tendency to say, "Well, their incentive is gone, they won't produce anymore." But the opportunities that they would have exploited are still around for other people to exploit. And once people hit their limit, if they really enjoy what they're doing, they can keep on doing it just for the fun of it. If they don't enjoy what they're doing, well then, hurray. Here's an opportunity to become something else. I've hit my limit — I can go write a book or I can be a gardener or I can try to be a professional tennis player and I won't starve because I get beat all the time or whatever.

David Cayley
The third aspect was limited births. Incomes, births, depletion of resources. This is probably the hardest one. It's certainly the most difficult for me to contemplate. Can you say, first of all, what your initial proposal on this was and then what changes you've gone thorough in relation to it?

Herman Daly
The original proposal was something that actually Kenneth Boulding had first suggested, an orphan brainchild of his that I adopted.

David Cayley
You mean he'd already been driven out of town over it?

Herman Daly
Yes. I love Kenneth Boulding. He's one of the people I've learned most from, but you know he first proposed it by saying, "In all seriousness, I believe that," and then later on he referred to it as, "A few years ago I somewhat jokingly suggested that ..." But the idea was, if you can see that reproduction has to be limited, then let's create a new right to reproduce, a legal right to reproduce, and let's distribute that right equally, on the basis of total equality. One person, one right, or each woman two rights. There are various ways you could do it. And then not everybody wants to reproduce, not everybody can reproduce. Those who don't could give, trade, sell their right to somebody who wants more than two children and can afford to buy it or can finagle you into giving it to them. Many people react with horror to this idea and they say, "Oh, the rich will have an advantage. The rich will have an advantage." Yes, that's true. The rich always have an advantage; that's the whole point of being rich. The rich buy Cadillacs and the poor can't; the rich get access to blood when they need it for operations, the poor don't. This is true; and if we don't want that to happen, then the way to do that is to limit the total advantages of the rich by the other institution, which is the limits to inequality — the minimum and the maximum. We say, "We're going to take care of that with a minimum income so that people will not be disadvantaged beyond some point." In addition, this idea does not involve buying and

selling children. It proposes a legal right to reproduce. If more children are born to richer rather than poorer parents, then there's something to be said for that. That's a benefit to the children. It tends to equalize the per capita distribution of income if that happens.

Now the problems come, of course, in enforcement. How would you enforce this right? Well, any sort of population-control scheme is going to face the problem of enforcement and I don't know really what the best ways are. You want ways which do not penalize the innocent child. Unfortunately, in China, some of the means of enforcement fall rather heavily on the children and indeed the whole family. Food rations are not increased, or are limited, and so on. Consequently, I'm not sure what would be the best forms of enforcement or punishment.

People look upon this proposal as a restriction of freedom. Sure it is. But if you go back and you read the classic defense of freedom, John Stuart Mill's *On Liberty*, you find that he makes a specific case for the right of the state to see to it that a country is not overpopulated — by laws which delay marriage until a couple is able to support children, or by various things like this. So there's a long and solid tradition of arguing that there is a collective interest in limiting a nation's population. Now in the nineteenth century it was commonly the practice not to do this, but on the other hand, you see, there was no welfare state. So the rule then was to let the unfortunate offspring starve and so that was a pretty effective way of dealing with overpopulation. It was a very cruel way and not very many people advocate that today. But if we want to move away from that, if we want to adopt a rule that unfortunate offspring are not going to be allowed to starve, they're going to be taken care of by society, then we have a correlative obligation to see to it that there are not so many unfortunate offspring that we can't deliver on that obligation. This is a big problem and we still need to keep plugging away at it.

David Cayley
Well, I can't say that there isn't a problem. Obviously your proposals are unpalatable; everybody else's proposals are unpalatable. What's happened in China, for example, seems pretty grim. And yet, it's hard to say, "Well, just let it go. Population is self-limiting at some level."

Herman Daly

Yes, it certainly is self-limiting. It's self-limiting in Malthusian terms. Malthus called misery, starvation and vice — he considered birth control to be a vice — positive checks. The neo-Malthusians said, "No, birth control is not vice; birth control is prudence." So I'm very much a neo-Malthusian in that sense. We can exercise some foresight and social planning on the issue of birth control.

The Chinese deserve a prize in gratitude from all mankind for having been the first society that seriously tried to deal with the problem. They were driven to it. For many years they kept repeating the old Marxist line that we have to protect the people from capitalism, not capitalism from the people, the more people the better, and so on. Well, they've backed off from that. Now, the other problem there is not just a matter of population, it's also a matter of per capita consumption. What's really limited is the aggregate throughput from nature, the total flow from nature through the economy, back to nature, at some sustainable level. That's equal to population times per capita resource consumption. We can operate on either of those variables. In rich countries we could say, "Oh, it's good to have a lot of people, let's just lower per capita consumption. We don't need all this stuff, let's have more people." We could do that. In poor countries that's a lot more difficult to do because they're much closer to the minimum necessary, so their only alternative is to work on the people factor of the equation and not the per capita consumption side.

David Cayley

What is the point, do you think, of making proposals as radical as yours in the present circumstances, where obviously they're not going to be immediately adopted by anyone?

Herman Daly

They won't be immediately adopted, but I guess the reason is that we think that the present circumstances won't be maintained. Things are going to get a lot worse and then these costs of growth will become so prevalent that everyone can see them. And already people are far ahead of the politicians. People are much more willing to accept leadership and recognition of these constraints than the politicians are. So at some point, after there's been a big disaster for environmental reasons, then we will get serious and want to reconstruct and do things differently. At least then there will be something on the table to start with. The discussion won't have to start from scratch. In a sense, neither John nor I would be very comfortable if we were suddenly made dictators and told to put everything into practice. We feel that we ought to have to go through the gauntlet of convincing people because that's a kind of verification, because we're like other people and these same arguments that convinced us ought to convince other reasonable people. If they don't, then maybe there's something wrong with our arguments. So we have a certain amount of faith that reason and argument and persuasion are effective and will prevail. To the extent that they're not workable, we're quite willing to re-examine our own views. Maybe we're wrong.

David Cayley
Do you fear that as the era of diseconomic growth continues, this actually has a disintegrating effect on society, that a moral disintegration is actually taking place? Do you fear that you're losing precisely what you need to reconstruct the world along the lines you've envisioned?

Herman Daly
Yes, that is the biggest danger. What we need is to build on the remnants of community that exist in order to enact these limits; and if the very system that denies limits is destroying the community, which is necessary to impose the limits, then we're in a real bind. That's a real impasse. So our hope is that there's still enough community left where we can begin to consolidate and build on it, before we tear things up too much. You've really, I think, put your finger on what is a real danger, that the corrosive effect on community is doubly bad because it's precisely community that you need in order to limit this increasing corrosive effect.

David Cayley
At the end of the 1960s and the beginning of the 1970s, there were a number of books that created a mood that you might call ecological pessimism, not optimistic about the human pro-

spect. One of the essays that you anthologized in one of your earlier volumes was William Ophul's, *Leviathan or Oblivion?*, which puts it pretty starkly. Now at that time you were not exactly participating in that mood. You were more concerned to make proposals. It's almost twenty years later. What is your mood today?

Herman Daly
That's an interesting question. All during the years in which I was teaching a lot of this stuff, more than once I had students come to me and say, "Oh, Professor Daly, this is all so pessimistic. I'm going to drop your course. It's just ruining my life. There are too many things to worry about." And I had to take that seriously because these were youngsters and many of them were not well-equipped to deal with really pessimistic, serious things. I'd like to make a distinction between pessimism and optimism on the one hand and hope and despair on the other. I would say whether you're a pessimist or an optimist, turns on a kind of betting man's rational expectation about the way things are likely to turn out. In that sense, I have to say I'm a pessimist. On the other hand, hope and despair are existential attitudes that you impose on the world from your own being or you just say, "In spite of the fact that I am pessimistic, I will be hopeful because it is a sin to despair and hope is the proper attitude, so I will be hopeful, and I hope that my calculations are wrong. I hope I lose the bet. I will set about doing things to try to see to it that I do lose that bet." So that's a kind of way of squaring that circle or at least living with both things.

*** *** ***

Herman Daly offers a series of practical proposals for achieving a steady-state economy. He's quite prepared to admit that they might not be the right proposals, but he wants at least to propose the steady state as a practical problem which can, in principle, be solved. This has the great virtue of bringing discussion down to earth. It lets us see what it might mean to put a fence around the market, or limit the flow of natural resources into the economy. His approach is careful and discriminating. He wants to retain what he finds compelling in modern economics, but at the same time to severely limit its scope in the interests of nature and community. Daly's approach is very much a middle way. He offers strong and unpalatable medicine, but he does want to address the existing institutions of development in a language they can potentially grasp and with proposals they can potentially adopt. This puts him somewhere between Wolfgang Sachs, who wants to simply eliminate development as a framework, and, say, the Brundtland Commission which wants growth and conservation at once. If there is to be reform, Daly's proposals, as I said at the beginning, are probably its best hope.

15

The Ambivalence of Ecology

To call the time in which we live the age of ecology is to say no more than is obvious. Air and waters, forests and grasslands, climates and soils are all being degraded so rapidly that even those who approach doomsday scenarios sceptically must wonder if some of these trends will not soon become irreversible. Much irreparable damage has already been done. Some ecological restoration is possible over long time periods, but no one can remake extinct species or rebuild the cultures that will die along with the places they depend on.

There is also widespread agreement on the inadequacy of the classical economics that treated nature as a practically limitless source of free goods, and an infinitely capacious sink for wastes. The "vandal ideology," as political theorist Victor Ferkiss once called this view, has had its day. A new consensus is emerging around the idea of sustainability. But this is where the difficulties begin rather than end. Does the concept imply a new kind of environmentally managed growth, as in the Brundtland Commission's vision of sustainable development? Does it imply the steady state envisaged by Herman Daly? Or does it imply shrinkage in the scale of economic activity? Sustainability can be a reductive one-dimensional concept, which looks only at how to safeguard the flow of resources to markets and ignores other aspects of the being of those "resources." Or it can be, in the hands of a thinker like David Brooks, "a quality of life" concept that refers not to physical growth but to "realization of potential." Sustainability can mean a kind of environmental triage, an attempt to calculate how much ecological damage is endurable, or it can mean stepping back from the edge. It can be a tool of global management or of local sovereignty. "The right means in the hands of the wrong man work in the wrong way," says an old Chinese yoga text. "The wrong means in the hands of the right man work in the right way." A lot depends on who's talking.

Ecology was once hailed as a map for remaking the world. Paul Shepard called it "the subversive science"; Paul Sears suggested that it might be "an instrument for the long-run welfare of mankind"; William Irwin Thompson called it "the new political science." This enthusiasm now needs to be tempered. So long as ecology named only an academic science and a protest movement, it was possible for the ambivalence of ecology to go unremarked. The fact that ecology signified an ideological complex containing both scientific and anti-scientific perspectives, modern and anti-modern aspirations, technocratic and romantic conceptions of nature was, in a sense, its strength. It gave the environmental movement a kind of hybrid vigour, as Wolfgang Sachs has observed. But, as environmental questions enter and eventually permeate politics, ecology has to be considered as a form of rule and reduced to a definite political direction. Liberalism, communism, and populism have all made this transition from utopian protest to ruling ideology, and in the process have revealed a character unanticipated by their intellectual ancestors. It is now ecology's turn; and, as the moment approaches, a gulf opens between the disparate perspectives that have constituted the environmental movement so far.

Many of these disparate perspectives have appeared in these pages. Their variety resists reductive explanatory schemes, but I do think it is possible to discern a dividing line between two fundamentally different discourses. In my introduction I drew this line between hubris and humility, between those who think that ecology mandates planetary management, and those who think that management on such a scale is neither possible not desirable. Here I want to try and make the same distinction between what I will tentatively call globalism; and, even more tentatively, conservatism.

Globalism takes the planet as the indivisible basis of both thought and action. It is a complete and fundamentally new world view, and, accordingly, it has both its mystical and its managerial sides. Systems analysts like William Clark, whom I quoted in my introduction, speak of "managing the planet." Theologians like Hans Kung speak of a new "ethic of global responsibility." The centre of globalism's concern is life, a word which has broken free of its anchoring in individual persons to become an end in itself. In Darwin's evolutionary theory, earth's flora and fauna evolved by trial and error against a relatively stable background. The niches for which they competed were given. In Lovelock and Margulis's Gaia hypothesis, it is life itself that evolves. "The atmosphere," Lovelock says, "is not an environment for life. It's something that life has made as an environment for itself ... Environment and life ... constitute a single cybernetic system." For William Irwin Thompson this implies "the end of nature" as a stable framework from which human beings can derive any standard of goodness, beauty or permanence. Nature, as Gaia, is a fluid system in constant interaction, which can yield no standard beyond its own survival. The earth becomes a magic bubble in space, self-subsistent, self-evolving, without up or down, inside or outside, its dynamic principle neither God, nor man but life.

Globalism is undergirded by what Thompson calls "the new biology" in his anthology *Gaia: A Way of Knowing*. This new science deals with "autopoetic systems," systems like Lovelock's Gaia, which literally make themselves. Classical science spoke of the laws of nature. The new biology rejects the idea of immutable laws. It sees order emerging out of chaos according to principles immanent within the system. The principles change as the system changes. There is nothing outside the system that can provide a stable platform for either observation or criticism.

"We have changed our point of view," Francisco Varela says in an essay in *Gaia: A Way of Knowing*, "from an externally instructed unit with an independent environment linked to a privileged observer, to an autonomous unit with an environment whose features are inseparable from the history of coupling with that unit, and thus no privileged perspective." What this means, simply put, is that human beings are as indistinguishable from their world, as their world is from them. We are emergent properties of life, and our boundaries blur into life's continuum. This is a systems view, which sees all truth as immanent within systems — there is no outside from which to constitute "a privileged perspective."

Gregory Bateson, writing in the same volume, translates Varela's insight into theological language by saying that his philosophy gets rid of "the separation between God and his creation — that sort of thing doesn't exist anymore." Without separation, of course, God is creation, and, more importantly, creation is God. God is simply the systemic ensemble of what evolves, which is to say, in effect, that life is God. Varela regards this epistemological manoeuvre as taking him past "a split Cartesianism to a world of no distance by mutual inter-definition." I would say rather that he has accomplished the characteristically post-modern *reductio ad absurdum* of Cartesianism. He has overcome the meaninglessness of matter by assimilating it within mind. Either way, all transcendent terms like *reality*, *truth*, or *God* are dissolved, and the system within which we are imbedded becomes the limit of understanding, as well as the proper object of management, study, and devotion.

These remarks lead into philosophical and theological issues that are beyond the scope of this essay, but it is important to at least note here what some of the implications of Gaia as "a way of knowing" are. I don't mean to suggest that every bureaucrat at the World Bank now shares Thompson's and Varela's philosophy of science. Globalism is just a provisional pigeonhole, not a uniform and self-conscious philosophy. And yet, I do think there are telling analogies between the science of cybernetic systems, which has produced the idea of Gaia, and new managerial modes that seek to administer it responsibly. For, if the earth comprises a single indivisible ecology, and this ecology is now threatened by the isolated and uncoordinated actions of individual societies, then government must also eventually become global. Once, the elimination of war was the mainspring of efforts at internationalism; now survival provides an even more compelling rationale. What Wolfgang Sachs calls ecocracy, or rule by ecologists, is certainly in its infancy, but the outlines are already clear. The invention of international development after World War II was already a harbinger of a much more homogeneous world; the new ecological imperium is likely to continue this trend with each country's rights and responsibilities adjudicated in terms of global "carrying capacity."

Globalism erases boundaries and dissolves cultures, insofar as they are defined by their boundaries. Cultures grow in particular places. They assert that these places are the centre of everything. They limit themselves according to a certain conception of the good, and they clothe this conception in forms that establish the reality of identity and otherness for that people. A planetary culture which is anything more than a euphemism for Western hegemony seems in this sense to be a contradiction in terms. As a species, we have nothing in common but our biology. What we reach for beyond our natural being is always expressed in specific histories and rituals, gestures and words that cannot be reduced to a general form. Boundaries may be arbitrary; but, without them, we are swept into the limitless world of process and system. This is William Irwin Thompson's world of "enclouded selves" and "distributive lattices," a world that is everywhere and nowhere at once. It is a world without a ground, in which the human cannot be clearly distinguished from its surroundings, and life in general acquires the privilege that once pertained to the particular delimited lives of persons.

"Think globally, act locally" has long been a slogan of the environmental movement. It implies that people can think in global terms, while still preserving an undivided sense of belonging to a particular place. But can this actually be done? Certainly, people in the past have lived in empires in which cosmopolitan and local identities overlapped. But globalism requires something novel and unprecedented: allegiance to an ecological construct. One can see the inevitability of this fate, and still wonder if it won't tend to dissolve all definite and delimited forms of inherited culture and leave us floating in an abstract unity, which Hegel once satirized as "a night in which all cows are black."

Doubts about globalism have begun to coalesce in the last few years into a conservative reaction. American poet and essayist Wendell Berry became probably the best-known exponent of this reaction a couple of years ago when he published an essay in *Harper's* magazine called "The Futility of Global Thinking," but there are many others. Calling them conservatives at a time when the word is often used to describe a species of atavistic liberalism is problematic, of course; but I use the word here in its more literal sense.

Conservatives tend to see managerial ecology as the final triumph of the economic and utilitarian ideology of the modern West. This absorption of ecology into economics is very evident in the report of the Brundtland Commission, which first put the term *sustainable development* into wide circulation in 1987. Economy and environment, say the authors of the report, are now interdependent variables within a single system. Change in one immediately brings change in the other. Not only does the health of the economy depend on the health of the environment, but the health of the environment also depends on the health of the economy. It is when the economy cannot provide for people that they are forced to encroach on dwindling forests and farm on marginal, easily eroded lands. The report argues, therefore, that the way to preserve the environment is to create "a new era of economic growth," only this time, of course, sustainable growth. This prescription extends the hegemony of economic thinking into every cranny of the natural world by making the environment depend on the economy as well as the economy on the environment. Nature will be allowed to exist outside of parks only where it can declare good and sufficient reasons for its existence before the bar of economy. A commander responsible for razing a village during the Vietnam War offered the subsequently notorious rationale that the town had to be destroyed in order to save it. In this case, nature has to be economized in order to save it.

William Petty, who was one of the ancestors of modern economics, described his study as "political arithmetick," a term that makes clear the extent to which economics is defined by calculation. But calculation is the enemy of all forms of loyalty, solidarity, and friendship, which depend on our obeying our commitment and not our interest. For this reason, even modern people have kept certain parts of their lives free of economic considerations and have done certain things simply because they were good to do. The current attempt to live permanently at the panicky edge of "sustainability" threatens to make the régime of economic calculation complete and universal. In some circles, environmental hygiene is already an obsession, with every gesture weighed for its environmental impact. It may soon be an anti-social act to release carbon dioxide into the atmosphere by lighting a fire. But this new hygiene is not concerned with nature as such. It is concerned with calculated effects, and environmental tolerances. It grows out of economic considerations rather than immediate contact with real places. What Marx called the commodity form has simply been extended to the natural world.

Part of what the conservatives want to defend is people's traditional ways of making do within the limitations of their natural surroundings. All enduring cultures have known how to live where they are, but they have done so in an integral way without abstracting from their way of life a separate science of ecology. Development has interrupted this attunement, as Vandana Shiva points out in the case of the Green Revolution. The new age of sustainable development presents the further danger that enthusiastic experts will now try and assimilate traditional knowledge to a managerial mode that is profoundly antipathetic to culture. To call native Americans "the first ecologists," as is frequently done, is to say something profoundly false, while trying to say something that is undeniably true. To speak of an indigenous "science," say of agroforestry, runs into the same difficulty. Traditional stewardship of forests is certainly scientific in the sense that it involves a body of knowledge; but, insofar as this body of knowledge remains integral to a way of life, it has little else in common with that modern Western science, which began by dividing mind from the mechanical operation of nature, and has ended by reducing nature to pure cybernetic abstraction. Culture will not survive the tutelage of scientific ecology. The scandal of sustainable development, Vandana Shiva told the Citizens' Summit conference in 1988, is that peoples whose traditional ways of life have been invaded and undermined by development are now to be taught ecology.

Wolfgang Sachs in many ways typifies the conservative critique of managerial ecology. At the heart of his thinking is the venerable idea that politics should involve a question about "the good life"; and, in the early stirrings of the Green movement in Germany in the 1970s, he perceived a question of this kind. He was concerned about how to live in a spirit of both decency and celebration within culturally defined limits and not just with prudential and utilitarian calculations of environmental risks. Nuclear power plants were bad not just because they presented unacceptable risks, but because they forced societies to engage in risk assessment in the first place. In Sachs's vision freedom and renunciation supported each other: by reducing the weight of its demands on the world, and moving back from the edge of calculated sustainability, a society could loosen the grip of purely economic considerations and pursue other ends.

By the end of the 1980s, Sachs had seen survival replace the good life as the pre-eminent objective of environmentalism. And in the calls he was hearing for securing survival and managing the planet, he saw "a grand operation to render politics irrelevant." Calculating the limits of sustainability and then "steering a precarious course along the edge" was to him the final triumph of technocracy. Instead of pursuing their particular vision of the good, societies would live under the surveillance of ecological experts, qualified in the global calibration of environmental impacts.

Ecology, for Sachs, poses not a technical but a moral question about how to live well. He does not suppose that tradition holds an exclusive, or even adequate answer to this question; but he also thinks that the attempt to make a political science of ecology without reference to tradition will eviscerate culture and reduce political considerations to biological ones. His answer is "global experimenting": abandon the universal and homogeneous ideology of development, and let the people again chart their own path toward the good life.

Another aspect of the conservative critique of managerial ecology is David Ehrenfeld's insistence on the inherent limits to knowledge. For him, the consequences of large-scale interventions in nature can never be fully foreseen or adequately controlled. In his analysis of the concept of "maximum sustained yield," he suggests that there are technical as well as moral flaws in the idea of sustainability. Consequently, he does not need to argue that intensive remedial management of nature is undesirable, since he believes on prior grounds that it is impossible. This leads him, like Sachs, to put his faith in renunciation and political de-centering rather than environmentally controlled growth.

The conservatives point to what John Livingstone once called "the paradox of environmentalism," which is that it betrays itself whenever it opens its mouth in public. Arguments for conservation inevitably take a utilitarian form. To speak of nature's value is implicitly to speak of nature's value for us. Even imponderables like beauty, when submitted to the "political arithmetick" of environmental assessment, are reduced to recreational or aesthetic amenities. Arguments must take a logical form; but, as Livingstone says, "the existence of the whooping crane is not logical." This is why arguments for "environmental ethics" always turn tautological and end up chasing their own tails. The rights with which we endow nature are purely our own invention. We live, Livingstone says, under "a metaphysical dome," trapped by our own anthropocentric assumptions. Environmentalism, finally, becomes its own enemy by arguing away exactly what it intends to conserve: the independent existence of nature.

What I have called a conservative approach to ecology emphasizes restraint, humility, and a recognition that the degradation of nature asks for a cultural rather than a technical response. It fears that the environmental crisis will spawn a new net of repressive and tutelary institutions, which will teach us how to live at the edge of nature's tolerance rather than stepping back from the edge. And it resists the abstract and monistic spirit of globalism in favour of more traditional conceptions of person and place.

Identifying globalism and conservatism as antithetical discourses is a way of clarifying discussion of the environment; but, obviously, it has its limitations. For every thinker like Sachs or Thompson who can be made to fit my categories, there are several others who will present some hybrid form. Between a world of small, local self-limiting cultures and a world of total planetary administration there are many intermediate positions. David Brooks and Herman Daly, for example, are thinkers who are prepared to use the rhetoric of sustainable development while at the same time trying to curb many of the pretensions of what I have called globalism.

In the end, what is important is that the implications of ecology be thought through as much as they can be before they confront us as unthought necessities. Choices are already limited. Global realities like ozone holes and the greenhouse effect already exist. But this does not mean that we can afford to ignore the fact that these uncanny new realities have the curious status of things we cannot sense but only know through the lens of scientific and technological expertise. The rhetorical figures of the conservatives also refer to doubtful and broken realities, things we can know through the lens of history or anthropology but which no longer exist around us. Living in new and unprecedented times, we have to pick our way cautiously through the dark. The questions that confront us will not go away, and they will not be solved. They are in a sense the constitution of the age of ecology, the tension within which the politics of the future will take shape.

Suggestions for Further Reading

The titles that follow are grouped so that they correspond to the sections in this book.

New Ideas in Ecology

Devall, Bill, and George Sessions, eds. *Deep Ecology.* Salt Lake City: Peregrine Smith Books, 1985.

Evernden, Neil. *The Natural Alien.* Toronto: University of Toronto Press, 1985.

Livingston, John. *The Fallacy of Wildlife Conservation.* Toronto: McClelland and Stewart, 1981.

Shepard, Paul. *Nature and Madness.* San Francisco: Sierra Club Books, 1983.

Citizens at the Summit

Adams, Patricia, and Lawrence Solomon. *In the Name of Progress: The Underside of Foreign Aid.* Toronto: Energy Probe Research Foundation, 1985.

———. *Odious Debts: Loose Lending, Corruption, and the Third World's Environmental Legacy.* Toronto: Earthscan, 1991.

Shiva, Vandana. *Staying Alive: Women, Ecology, and Development.* London: Zed Books Ltd., 1988.

From Commons to Catastrophe

Anderson, Robert S., and Walter Huber. *The Hour of the Fox: Tropical Forests, the World Bank, and Indigenous Peoples in Central India.* Seattle: University of Washington Press, 1988.

Banford, Sue, and Oriel Glock. *The Last Frontier: Fighting Over Land in the Amazon.* London: Zed Books Ltd., 1985.

Caufield, Catherine. *In the Rainforest.* London: Picador, 1986.

Fernandes, Walter, ed. *Forests, Environment and People: Ecological Values and Social Costs.* New Delhi: Indian Social Institute, 1983.

────── and Sharad Kulkarni, eds. *Towards a New Forest Policy: Peoples, Rights and Environmental Needs.* New Delhi: Indian Social Institute, 1983.

Gillis, Malcolm, and Robert Repetto. *Public Policies and the Misuse of Forest Resources.* New York: Cambridge University Press, 1988. (A digest of this book called *The Forest and the Trees* is available from the World Resources Institute, 1735 New York Avenue, NW, Washington, DC 20006.)

Goodland, R. J. A., and H. S. Irwin. *Amazon Jungle: Green Hell to Red Desert.* New York: Elsevier Scientific Publishing Co., 1975.

Gradwohl, Judith, and Russell Greenberg. *Saving the Tropical Forests.* London: Earthscan Publications, 1988.

Myers, Norman. *The Primary Source: Tropical Forests and Our Future.* New York: W. W. Norton and Co., 1984.

Nectoux, Francois, and Yoichi Kurada. *Timber from the South Seas: An Analysis of Japan's Tropical Timber Trade and Its Environmental Impact.* Gland, Switz.: A World Wildlife Fund International Publication, 1989.

Posey, Darrell, et. al. "Enthnoecology, as Applied Anthropology in Amazonian Development." *Human Organization* 43.2 (Summer 1984).

Schultes, R. E. "Conservation Looks to the Medicine Man." *Social Pharmacology* 2.1 (1988): 83–91.

────── "Primitive Plant Lore and Modern Conservation." *Orion Nature Quarterly* 7.3 (1988): 8–15.

────── "From Witch Doctor to Modern Medicine: Searching the American Tropics for Potentially New Medicinal Plants." *Arnoldia*, 32.5 (September 1972).

The Age of Ecology
Bilsky, Lister J., ed. *Historical Ecology: Essays on Environment and Social Change.* Point Washington, New York: Kennikat Press, 1980.
Bramwell, Anna. *Ecology in the Twentieth Century: A History.* New Haven: Yale University Press, 1989.
Davis, Donald Edward. *Ecophilosophy: A Field Guide to the Literature.* San Pedro: R & E Miles, 1989.
Ehrenfeld, David. *The Arrogance of Humanism.* New York: Oxford University Press, 1978.
Garb, Jerome Yaakov. "The Use and Misuse of the Whole Earth Image." *Whole Earth Review* March 1988.
Kohak, Erazim. *The Embers and the Stars.* Chicago: University of Chicago, 1984.
Leopold, Aldo. *A Sand County Almanac.* New York: Oxford University Press, 1949.
Livingston, John. *The Fallacy of Wildlife Conservation.* Toronto: McClelland and Stewart, 1981.
Lovelock, James. *The Ages of Gaia.* Toronto: W. W. Norton, 1988.
McKibben, Bill. *The End of Nature.* New York: Random House, 1989.
Paehlke, Robert. *Environmentalism and the Future of Progressive Politics.* New Haven: Yale University Press, 1989.
Plant, Judith, ed. *Healing the Wounds: The Promise of Ecofeminism.* Toronto: Between the Lines, 1989.
Rothenberg, David, with Arne Naess. *Ecology, Community and Lifestyle.* New York: Cambridge University Press, 1989.
Ryle, Martin. *Ecology and Socialism.* London: Radius, 1988.
Shiva, Vandana. *Staying Alive: Women, Ecology, and Development.* London: Zed Books Ltd., 1988.
Thompson, W. I., ed. *Gaia: A Way of Knowing.* Great Barrington, Mass.: Inner Traditions/Lindisfarne Press, 1987.
Todd, John. "Adventures of an Applied Ecologist." *Whole Earth Review* 62 (Spring 1989).
―――― and Nancy Jack Todd. *Tomorrow Is Our Permanent Address.* New York: Harper & Row, 1980.
Worster, Donald. *Nature's Economy: A History of Ecological Ideas.* New York: Cambridge University Press, 1977.

Redefining Development

Adams, Patricia, and Lawrence Solomon. *In the Name of Progress: The Underside of Foreign Aid*. Toronto: Energy Probe Research Foundation, 1985.

———. *Odious Debts: Loose Lending, Corruption, and the Third World's Environmental Legacy*. Toronto: Earthscan, 1991.

Daly, Herman, and John Cobb, Jr. *For the Common Good: Redirecting the Economy toward Community, the Environment, and a Sustainable Future*. Boston: Beacon Press, 1989.

———. *The Steady State Economics: With New Essays*. 2nd ed. Washington: Island Press, 1991.

Sachs, Wolfgang, ed. *Development: A Polemical Dictionary*. London: Zed Books Ltd. (forthcoming).

Vachon, Robert et al. eds. " Archeology of the Development Idea: Six Essays by Wolfgang Sachs." *Interculture*. 109 (Fall 1990).